4th Workshop on Directional Detection of Dark Matter

(CYGNUS 2013)

Journal of Physics: Conference Series Volume 469

Toyama, Japan
10 - 12 June 2013

ISBN: 978-1-62993-937-7
ISSN: 1742-6588

Printed from e-media with permission by:

Curran Associates, Inc.
57 Morehouse Lane
Red Hook, NY 12571

CURRAN ASSOCIATES INC.
proceedings
.com

Some format issues inherent in the e-media version may also appear in this print version.

Printed by Curran Associates, Inc. (2014)

Institute of Physics
Dirac House, Temple Back
Bristol BS1 6BE UK

Phone: 44 1 17 929 7481
Fax: 44 1 17 920 0979

techtracking@iop.org

Additional copies of this publication are available from:

Curran Associates, Inc.
57 Morehouse Lane
Red Hook, NY 12571 USA
Phone: 845-758-0400
Fax: 845-758-2634
Email: curran@proceedings.com
Web: www.proceedings.com

TABLE OF CONTENTS

Journal of Physics: Conference Series **469** (2013) 012001 doi:10.1088/1742-6596/469/1/012001

Updates from the Dark Matter Time Projection Chamber Group (DMTPC)

James B. R. Battat [for the DMTPC Collaboration]

Wellesley College, Physics Department, 106 Central Street, Wellesley, MA 02481, USA

E-mail: jbattat@wellesley.edu

Abstract. The Dark Matter Time Projection Chamber (DMTPC) collaboration has developed a series of gas-based detectors with the goal of detecting the directional anisotropy of dark-matter-induced nuclear recoils. Here, we report on recent progress from the DMPTC group, focusing on the surface operation of the 4shooter detector.

1. Introduction

In early prototype detectors, the DMTPC collaboration demonstrated successful track reconstruction, including head/tail sensitivity, with CCD-based TPCs [6, 5]. Since then, the group has built and operated a 10-liter-volume detector, called the 10L. At the time, a surface run of that detector produced the most sensitive directional limit on the spin-dependent WIMP-proton cross-section [1]. Since then, the NEWAGE collaboration has matched this result (see NEWAGE contribution in this proceedings), and the DRIFT group have published a more senstive limit [3]. The 10L detector has been relocated to an underground lab facility at the Waste Isolation Pilot Plant (WIPP) near Carlsbad, New Mexico (USA), and an underground run is currently underway.

In parallel with the 10L work, the DMTPC group designed and built a new detector called the 4shooter. The 4shooter is a modest scale-up of the 10L technology (by a factor of two in volume), with the goal of evaluating technologies and techniques that would be required for a cubic-meter DMTPC detector.[1] In particular, the 4shooter uses four CCD cameras to produce a mosaic image of the TPC amplification region. Also, in addition to the CCD readout, the 4shooter has both photomultiplier tube (PMT) and charge readout channels, with the goal of enhancing background rejection and 3D tracking capability.

In this work, we present recent results from the operation of the 4shooter detector at a surface laboratory at MIT (in Cambridge, MA, USA). In particular, we have characterized the gas gain, and calibrated the energy scale of the CCD and charge readout channels. In addition, we describe initial results on gamma background rejection, CCD track reconstruction techniques, and the head/tail reconstruction capability for low-energy (50–400 keV) alpha tracks.

[1] The cubic meter detector, named DMTPCino, has been funded for construction by NSF and DoE, and is currently under design by the DMTPC group.

Journal of Physics: Conference Series **469** (2013) 012001 doi:10.1088/1742-6596/469/1/012001

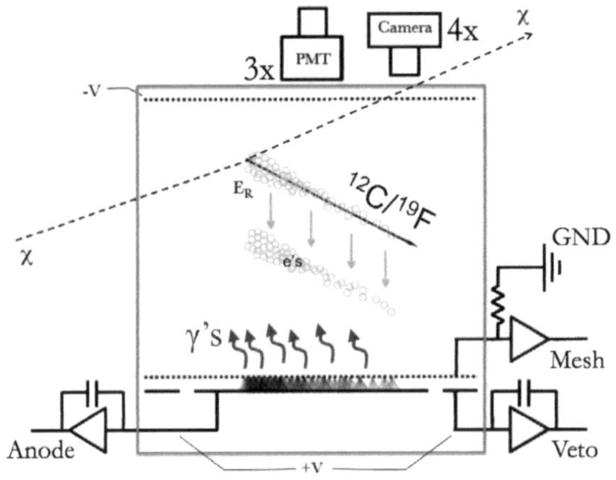

Figure 1. Schematic of the 4shooter detector showing a side-view of the TPC with mesh cathode ($-V$), mesh ground plane, and solid anode and veto electrodes ($+V$). A recoiling C or F nucleus ionizes the surrounding gas. Proportional amplification and scintillation light is produced in the mesh-anode gap (the amplification region). The charge signal is measured by the charge-integrating amplifiers on the anode and veto channels, as well as the high-bandwidth current amplifier on the mesh. The scintillation light is recorded by four CCD cameras and three photomultiplier tubes (PMT).

2. 4shooter Detector Hardware

As shown in Figure 1, the 4shooter detector is a time-projection chamber apparatus, housed within a vacuum chamber, with both charge and optical readout. The detector is filled with low-pressure CF_4 gas (45–75 Torr). The 26.7-cm tall drift region couples to a 420 μm tall amplification region through a stainless steel mesh that is common to both. Copper field shaping rings with 30.7 cm inner diameter provide a uniform drift electric field. The amplification region is a copper-clad G-10 plate with machined channels to define three separate electrodes: ground, anode and veto. The circular anode has a 29.2 cm diameter, and around the anode is an annular veto region of inner diameter 29.2 cm and outer diameter 30.7 cm. The veto channel is used for charge-based fiducialization in (x, y). Finally, the remainder of the copper-clad G-10 plate (exterior to the veto), comprises the ground electrode, to which the amplification region mesh is epoxied. The total active volume of the detector is then defined by the drift length and the outer diameter of the veto channel, while the nominal fiducial volume (18 L) is set by the drift length and the outer diameter of the anode.

3. Gas gain

Measurements of the gas gain as a function of anode voltage and gas pressure were made using an ^{55}Fe source of 5.9 keV x-rays. As shown in Figure 2, at 75 Torr CF_4, a gain of 10^5 was achieved (for an anode bias of 720 V). In standard operation, the detector is filled with 60 Torr CF_4 and the amplification region bias is 670 V. The resulting gas gain is 6.3×10^4.[2] Additionally, the gas gain as a function of time was found to degrade by 3% but then stabilize over a 24 hour period. This behaviour is well-documented in gas detectors, and may be caused by the formation of hydrocarbon polymers on the anode [7].

We also found that the measured gas gain decreased with the rate of the ^{55}Fe source, likely due to space-charge formation in the amplification region [2]. We attenuated the source until the measured gas gain plateaued. Measurements of the gas gain with the unattenuated source

[2] Note, that we have assumed a W-value of 34 eV for CF_4, though conflicting values appear in the literature (34 eV [10] and 54 eV [11]).

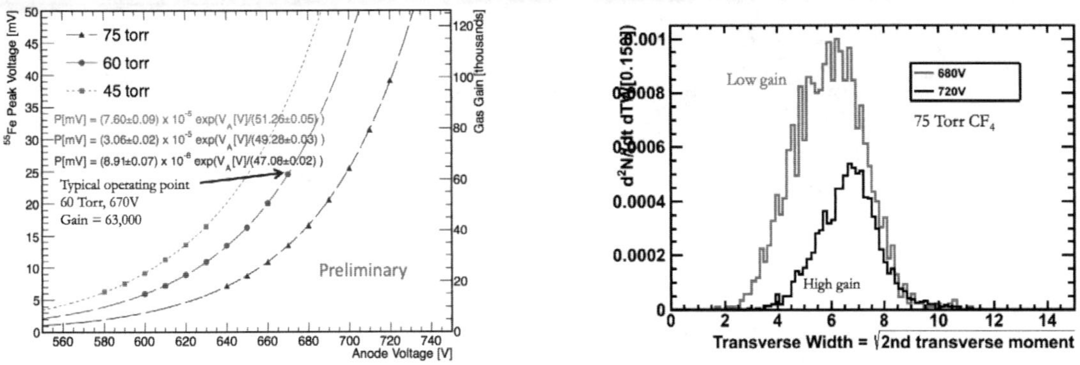

Figure 2. (Left): Gas gain as a function of anode voltage for three different CF_4 gas pressures. (Right): Distributions of transverse widths of neutron-induced nuclear recoils showing the absence of low-diffusion tracks at high detector gas gain, likely due to the Raether limit.

were 30% lower than with the attenuated source.

With the 4shooter detector, there is a high premium on gas gain, as the CCD acceptance is only 8×10^{-4}. But the trade-off of increasing the anode voltage to achieve a large gas gain is non-trivial. First, the rate of spontaneous discharges (sparks) increases. This reduces detector live-time and can lead to residual (ghost) images in subsequent CCD exposures. Second, the ionization density from a nuclear recoil, when amplified, may exceed 10^7 electrons (the so-called Raether limit), which can lead to event-triggered discharges. Particularly troublesome is the drift-distance dependence of the charge density (due to diffusion). For example, in Figure 2, we compare the transverse widths of neutron-induced nuclear recoil tracks in the CCD (the transverse width is correlated with the electron diffusion). We find that data at 75 Torr CF_4 and a high-gain setting (720V on the anode) shows an absence of low-diffusion tracks, while data at a lower gain (680V on the anode) shows the expected width distribution. A likely explanation is that the detector is blind to short-drift events at high gain because they trigger discharges in the amplification region due to Raether's criteria. During normal operations, we do not operate near the Raether limit.

4. CCD energy calibration

The energy scale of the CCDs is calibrated by comparing the longitudinal projection of ^{241}Am alpha particle tracks in each CCD field of view with Monte Carlo simulations (see Figure 3). The Monte Carlo parameters include the starting location of the source (tuned to match the mean range of the data), the electron diffusion (tuned to match the widths of the transverse projections of the tracks), the (negligible) gas quenching factor (using the parameterization of [5]), the CCD length scale (conversion from pixels to millimeters – obtained via independent measurements), the CCD noise per pixel (also obtained via independent measurements), as well as the desired parameter, the CCD conversion gain in units of ADU/keV$_{ee}$ (where ADU are the Arbitrary Digital Units reported by the CCD).

An example of the agreement between the data and Monte Carlo is shown in Figure 3. The resulting CCD energy calibration parameters for 60 Torr CF_4 and anode voltage of 670 V are 10.3, 18.4, 18.6 and 16.6 ADU/keV$_{ee}$ (all ± 0.2 ADU/keV$_{ee}$). The difference between the energy scale of the first CCD and the other three is due to a difference in the electron conversion gain (e^-/ADU).

Journal of Physics: Conference Series **469** (2013) 012001 doi:10.1088/1742-6596/469/1/012001

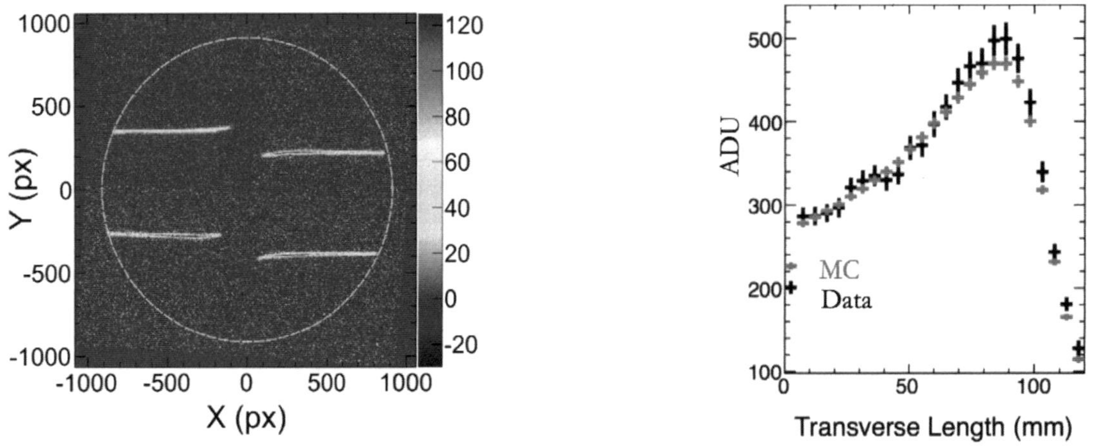

Figure 3. (Left): Mosaic image of the CCD energy calibration data showing a single alpha calibration source in the field of view of each CCD. The white dotted line marks the location of the inner diameter of the field-shaping rings onto which the alpha sources are mounted. (Right) The longitudinal projection of the average of many alpha tracks in a single CCD field of view for data (black) and Monte Carlo (red), showing good agreement between the two. The normalization of the Monte Carlo gives the CCD energy calibration in ADU/keV$_{ee}$.

5. Other results

The signal from the current amplifier attached to the amplification region mesh provides gamma background rejection via pulse shape discrimination. For example, with a similar detector (same amplification region geometry, but shorter drift region), the DMTPC group previously demonstrated a combined CCD/charge gamma rejection factor of 1.1×10^{-5} [8]. Although the background rejection factor has not yet been quantified on the 4shooter, the preliminary results suggest that we will achieve even better performance than in [8].

Additionally, a new approach to track reconstruction that employs a physical model for the nuclear stopping has been implemented. This approach not only leads to better track reconstruction accuracy, but also allows for the assignment of confidence/quality to each fit parameter, allowing for subsequent analyses to employ this confidence parameter. Two different fitting approaches have been developed: one-dimensional fits to the transverse and longitudinal projections of the 2D tracks in the CCD image, as well as a 2D fit to the original 2D track in the CCD image. Preliminary results of this work were presented recently at the TAUP 2013 conference [4].

Finally, we used an alpha source of known orientation situated outside of the drift region, oriented so that only the final 50–300 keV of the track penetrated through the cathode into the drift region. It was therefore possible to quantify the ability of the detector to correctly reconstruct the head-tail signature of low-energy ionization tracks. These tracks differ from the expected WIMP-induced nuclear recoils in two important ways. First, they are produced by alpha particles (not F recoils), so the track lengths are longer (optimistic), but the stopping (dE/dx) is smaller (pessimistic) than F recoils of the same energy. Second, the primary ionization is generated near the cathode, and so these events all travel the full 26.7 cm drift length of the detector and suffer the maximum diffusion (pessimistic). The promising initial results [9] show that the head-tail was measured correctly on 60% of tracks with energy between 60 and 80 keV. In the range of 100 to 120 keV, that fraction improved to 70%, and at 200 keV the fraction was 90%. Furthermore, the head-tail correct fraction is significantly increased with the use of the 2D track-fitting analysis described above [4].

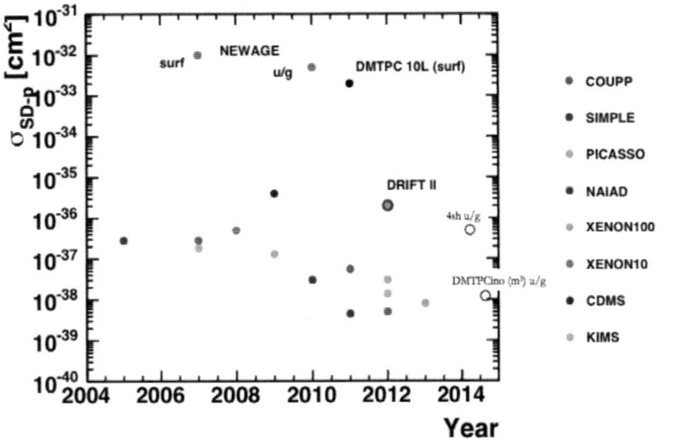

Figure 4. Limits on the spin-dependent WIMP-proton cross-section over time for both directional (DMTPC, DRIFT, NEWAGE) and non-directional detectors (listed in legend at right). The point labeled "4sh u/g" is the exclusion limit achievable if the 4shooter detector is operated background-free for one year with a 50 keV$_r$ threshold. The corresponding sensitivity of the cubic-meter DMTPCino is also shown.

6. Projected sensitivity

The evolution over time of spin-dependent WIMP-proton cross-section limits is presented in Figure 4. Over the past seven years, the limits have improved by two orders of magnitude. The sensitivity of the 4shooter and the cubic-meter DMTPCino detector are also shown.

References

[1] [DMTPC Collaboration] Ahlen, S. First dark matter search results from a surface run of the 10-L DMTPC directional dark matter detector. *Phys. Lett. B*, 695:124–129, January 2011.

[2] A. Bressan et al. High rate behavior and discharge limits in micro-pattern detectors. *Nucl. Instrum. Meth.*, A424:321–342, March 1999.

[3] E. Daw et al. Spin-dependent limits from the DRIFT-IId directional dark matter detector. *Astropart. Phys.*, 35:397–401, February 2012.

[4] [DMTPC Collaboration] Deaconu, C. Recent Updates from the Dark Matter Time Projection Chamber (DMTPC) Collaboration. In *13th International Conference on Topics in Astroparticle and Underground Physics (TAUP 2013)*, 2013.

[5] [DMTPC Collaboration] Dujmic, D. Charge amplification concepts for direction-sensitive dark matter detectors. *Astropart. Phys.*, 30:58–64, 2008.

[6] [DMTPC Collaboration] Dujmic, D. Observation of the "head-tail" effect in nuclear recoils of low-energy neutrons. *Nucl. Instrum. Meth.*, A592:123, 2008.

[7] J. Kadyk. Gaseous Detectors - Novel Tracking Detectors. `http://www-group.slac.stanford.edu/sluo/lectures/Detector-Lectures.html`, October 1998. Accessed 2013-09-22.

[8] [DMTPC Collaboration] Lopez, J. P. Background Rejection in the DMTPC Dark Matter Search Using Charge Signals. *Nucl. Instrum. Meth.*, A696:121–128, 2012.

[9] [DMTPC Collaboration] Lopez, J. P. Updates from the DMTPC Dark Matter Search. American Physical Society April 2013 meeting, April 2013.

[10] G. F. Reinking, L. G. Christophorou, and S. R. Hunter. Studies of total ionization in gases/mixtures of interest to pulsed power applications. *J. Appl. Phys.*, 60:499–508, July 1986.

[11] A. Sharma. Properties of some gas mixtures used in tracking detectors. *SLAC-JOURNAL-ICFA-16-3*, July 1998.

Journal of Physics: Conference Series **469** (2013) 012002 doi:10.1088/1742-6596/469/1/012002

MIMAC: MIcro-tpc MAtrix of Chambers for dark matter directional detection

D. Santos, G. Bosson, J.L. Bouly, O. Bourrion, Ch. Fourel, O. Guillaudin, J. Lamblin, F. Mayet, J.F. Muraz, J.P. Richer, Q. Riffard

LPSC, Universite Joseph Fourier Grenoble 1, CNRS/IN2P3, Institut Polytechnique de Grenoble

L. Lebreton, D. Maire

LMDN, IRSN Cadarache, 13115 Saint-Paul-Lez-Durance

J. Busto, J. Brunner, D. Fouchez

CPPM, Aix-Marseille Université, CNRS/IN2P3, Marseille, France

E-mail: `Daniel.Santos@lpsc.in2p3.fr`

Abstract.
Directional detection of non-baryonic Dark Matter is a promising search strategy for discriminating WIMP events from neutrons, the ultimate background for dark matter direct detection. This strategy requires both a precise measurement of the energy down to a few keV and 3D reconstruction of tracks down to a few mm. The MIMAC (MIcro-tpc MAtrix of Chambers) collaboration has developed in the last years an original prototype detector based on the direct coupling of large pixelized micromegas with a special developed fast self-triggered electronics showing the feasibility of a new generation of directional detectors. The first bi-chamber prototype has been installed at Modane, underground laboratory in June 2012. The first undergournd background events, the gain stability and calibration are shown. The first spectrum of nuclear recoils showing 3D tracks coming from the radon progeny is presented.

1. Introduction

Directional detection of Dark Matter is based on the fact that the solar system moves with respect to the center of our galaxy [1]. Taking into account the hypothesis of the existence of a galactic halo of DM formed by WIMPs (Weakly Interacting Particles), we can expect a privileged direction for the nuclear recoils in our detector, coming from elastic collision with those WIMPs [2][3].

The MIMAC (MIcro-tpc MAtrix of Chambers) detector project [4] tries to get these elusive events by a double detection: ionization and track, at low gas pressure with low mass target nuclei (H, ^{19}F). In order to have a significant cross section with such low mass nuclei, we explore the axial, spin dependent, interaction on odd nuclei. The very weak correlation between the neutralino-nucleon scalar cross section and the axial one, as it was shown in [5] and more recently in [6] makes this research, at the same time, complementary to the massive target experiments.

Journal of Physics: Conference Series **469** (2013) 012002 doi:10.1088/1742-6596/469/1/012002

2. MIMAC bi-chamber prototype

The MIMAC bi-chamber prototype is composed of two chambers sharing the same cathode being the module of the matrix devoted to show the ionization and track measurement performances needed to achieve the directional detection strategy.. The prototype active volume ($V \sim 5.8\,\mathrm{L}$) is filled with : $70\,\%\,CF_4 + 28\,\%\,CHF_3 + 2\,\%\,C_4H_{10}$ at a pressure of 50 mbar. The primary electron-ion pairs produced by a nuclear recoil in one chamber of the matrix are detected by driving the electrons to the grid of a bulk micromegas[7] and producing the avalanche in a very thin gap (256μm).

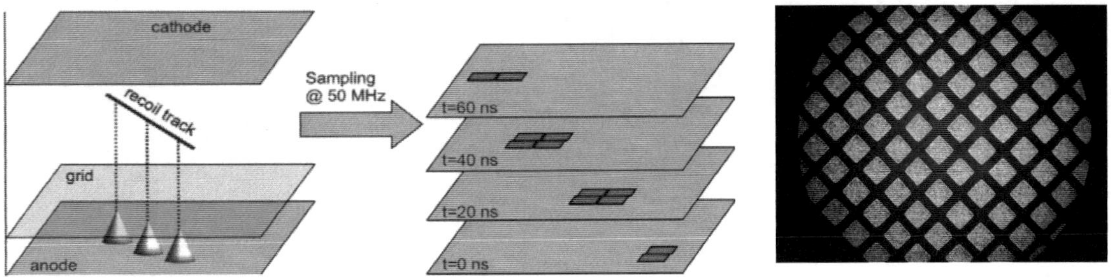

Figure 1. Left: Track reconstruction in MIMAC. The anode is read every 20 ns. The 3D track is reconstructed, from the consecutive number of images defining the event. Right: A picture of a small part of the pixellized micromegas designed by the IRFU (Saclay) team showing the 200 μm wide pixels [8].

As pictured on figure 1 left , the primary electrons are collected to the grid in the drift space (25 cm) and are multiplied by avalanche to the anode thus allowing to get information on X and Y coordinates. To access to the X and Y coordinates with enough spatial resolution, a bulk micromegas was developped [8] with a 10.8 cm by 10.8 cm active area, segmented in pixels 200 μm wide, defining a 424 μm pitch orthogonal structure shown on figure 1 right . In order to reconstruct the third dimension Z of a recoil track, the LPSC developed a self-triggered electronics able to perform the anode sampling at a frequency of 50 MHz. This includes a dedicated 64 channels ASIC [9] associated to a DAQ [10] specially developped for the project. The pixellized micromegas coupled to the MIMAC electronics running the 512 channels, is in fact the validation of the feasibility of a large TPC for directional detection.

3. MIMAC bi-chamber prototype at the Laboratoire Souterrain de Modane

The MIMAC bi-chamber prototype was installed at the Laboratoire Souterrain de Modane (LSM) in June 2012, see fig.2 left. By means of an X-ray generator, the detector is weekly calibrated by Cd (3.2 keV), Cr (5.4 keV), Fe (6.4 keV) , Cu (8.1 keV) and Pb (10.5 and 12.6 keV) fluorescence photons. Fig. 2 right shows the position of peaks fitted by a linear calibration in ADC channels as a function of time, highlighting the gain stability during the data taking period. The last two peaks from Pb X-rays suffer of an important statistical fluctuation due to the low cross section interaction at 50 mbar in our gas mixture. The gas circulation system including a buffer volume, an oxygen filter, a dry pump and a pressure regulator allows to keep the gas quality stable in a closed circuit. Indeed, the presence of impurities and O_2 must be controlled to prevent gain and energy resolution degradation.

Journal of Physics: Conference Series **469** (2013) 012002 doi:10.1088/1742-6596/469/1/012002

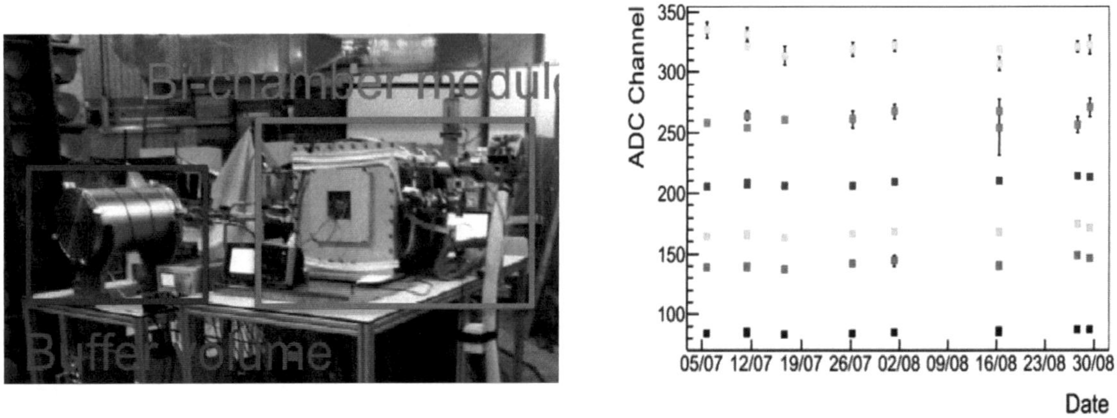

Figure 2. Left: The bi-chamber prototype at the Laboratoire Souterrain de Modane in June 2012. The bi-chamber module is identified in red and the buffer volume in blue. Right: The position of peaks of Cd (3.2 keV), Cr (5.4 keV), Fe (6.4 keV) , Cu (8.1 keV) and Pb (10.5 and 12.6 keV) fitted by a linear calibration in ADC channels as a function of time, highlighting the gain stability during the data taking period.

4. Preliminary analysis of the first months of data taking

The first available data set of the bi-chamber prototype was started on July 5th 2012 and finished on October 12th 2012. The measured total event rate of tracks with a 2 keV threshold was 5.6 ± 0.4 evts/min without any cuts. We performed a first data analysis that allowed us to obtain 3D-track events with an energy spectrum shown on Fig. 3 left. In this spectrum, we can clearly identify two peaks at 32.4 ± 0.1 and 44 ± 0.3 keV, and a flat distribution above 60 keV.

Fig. 3 right shows the event rate of α-particles (ADC saturation above 120 keV in red) and of events with energies between 20 and 60 keV selected by this analysis in blue. These event rates remained constant until October 3rd, when the gas circulation was stopped to isolate the bi-chamber from the gas circulation system. We can observe an exponential reduction of the event rates which fit with the ^{222}Rn half-life ($T_{1/2} = 3.8\,$days). It signs a pollution of the gas mixture by Radon isotopes contained in the circulation gas system due to a leak in the dry pump assuring the circulation of the gas from the buffer volume. The exponential decrease being dominated by the half-life of the ^{222}Rn, the contribution from the decay of ^{220}Rn may be neglected in a first order approximation.

The seven α-particles from the decay of ^{222}Rn and ^{220}Rn descendants, called Radons progeny, are emitted with kinetic energies E_{α}^{kin} ranging from 5.5 to 8.8 MeV. A simulation showed that these α-particles can pass through the 24 μm cathode to reach the other chamber even if the decay is produced in the gas volume. The simulated energy spectrum of the α-particles reaching the other chamber was flat from 0.3 to 120 keV , our ionization energy dynamic range. In addition, if the decays occur at the surfaces and if the α-particles are absorbed in the matter, only the recoils of the daughter nuclei will be detected. The recoils of the daughter nuclei from the Radon progeny have kinetic energies from 100 to 170 keV, see [14] and [15].

The difference between the kinetic energy of an ion E_{recoil}^{kin} and the energy released by ionization E_{recoil}^{ioni} in the active volume is determined by the Ionization Quenching Factor (IQF) defined as the ratio between the ionization energy released by a recoil and the ionization energy

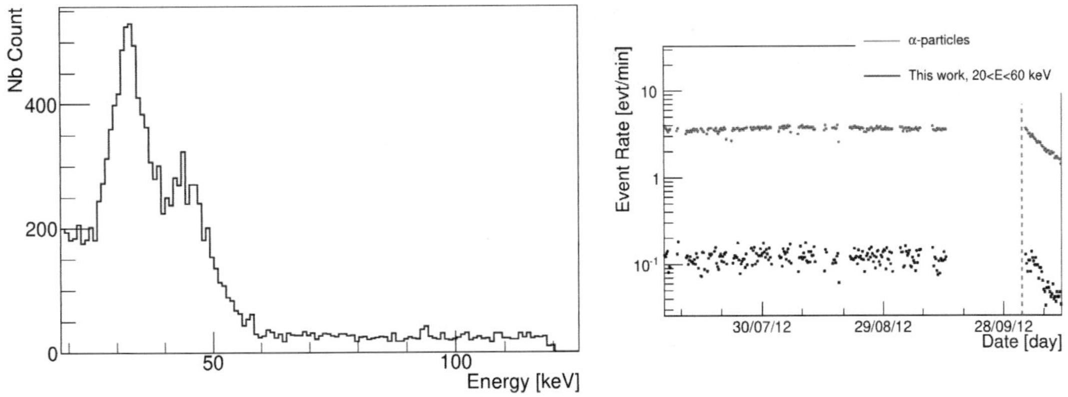

Figure 3. Left: The energy spectrum obtained after a preliminary analysis between 20 and 120 keV. Right: The event rates of α-particles (red dots) and the event rate of events selected by this analysis with energies between 20 and 60 keV (blue dots). The vertical dashed red line indicates the shutdown of the gas circulation system in October 3rd, 2012.

released by an electron at the same kinetic energy. Taking into account the *IQF* correction from SRIM [13], the recoils from ^{222}Rn and ^{220}Rn progeny would be detected with ionization energies from 38 to 70 keV. It was shown that SRIM overestimates the *IQF* roughly by 20 % at such recoil energies [11], [12], in the case of ^{4}He nuclei. There is no experimental data available for the quenching of ^{218}Po, ^{214}Pb or ^{210}Pb recoils in our gas mixture. This measurment may be one of the first evidence of the quenching in ionization for these heavy nuclei in a gas mixture. Taking into account the *IQF* estimated overestimation, the energies of the recoils from the ^{222}Rn progeny fit the spectrum. We can confidently conclude that the 3D-track events associated to the energy spectrum shown on Fig. 3left come from the ^{222}Rn progeny in the detector. The identification of the contribution of each nuclear recoil from the ^{222}Rn progeny will be presented in a future paper. As an illustration of the high quality of data obtained at Modane, we show on fig.4 a very interesting recoil event of 34 keV in ionization from the radon progeny discussed above.

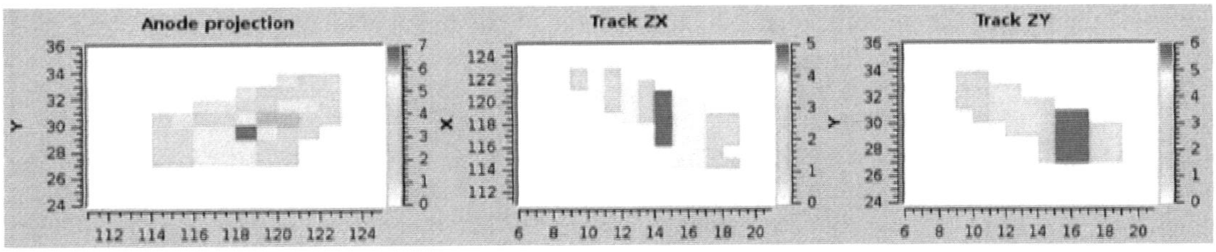

Figure 4. A 3D recoil track measured at 34 keV in ionization: On the left, the X-Y plane of the anode showing the intersections of the strips fired. On the centre, the X projection as a function of time, every 20 ns. On the right, the same but for the Y projection

For directional detection it is important to discriminate Radon progeny events from WIMP like events. This discrimination should be possible by using the time and spatial coincidence between the two chambers. For those events produced at the cathode surface, we expect one recoil event in one of the chambers and the α-particle through the cathode in the other one.

For this data taking period, the time synchronization of the chambers was not available. The detector was upgraded and the new version was intalled end of June 2013. The upgraded version has a three times bigger gas buffer volume, a thinner cathode (12 μm instead of 24 μm), a charcoal filter at -15 ^oC for radon filtering and the electronic synchronisation of the chambers. The evalutation of the rejection of the nuclear recoil events from the radon progeny by the coincidences between the two chambers will be performed in the next data taking period started in July 1st 2013.

The next step in the definition of a large TPC for directional detection is to build a demonstrator of $1\,m^3$. A preliminary mechanical design keeping the bi-chamber module as the elementary structure repeated with larger micromegas detectors (20 cm x 20 cm) is shown in fig.5.

Figure 5. The preliminary mechanical design of the demostrator of MIMAC -1 m^3 .

5. Conclusions

The MIMAC detector provides the ionization energy of a recoiling nucleus at low energies and the reconstruction of its 3D track. The bi-chamber prototype was installed at the LSM in June 2012. A preliminary analysis of the first data set allowed us to observe, for the first time, the 3D-track and energy of recoils from the ^{222}Rn progeny. The next step of the MIMAC project will be the development of the MIMAC $1\,m^3$. The 1 m^3 detector will be the demonstrator for a large TPC devoted to DM directional search.

6. References

[1] D. N. Spergel, Phys. Rev. D **37** (1988) 1353.
[2] J. Billard *et al.*, Phys. Rev. D **83** (2011) 075002.
[3] J. Billard *et al.*, Phys. Rev. D **85** (2012) 035006.
[4] D. Santos *et al.*, EAS Publ. Ser. **53** (2012) 25.
[5] E. Moulin, F. Mayet et D. Santos. Phys. Lett. B **614** (2005) 143-154.
[6] D. Albornoz Vsquez, G. Blanger, J. Billard and F. Mayet, Phys. Rev. D **85** (2012) 055023.
[7] I. Giomataris *et al.*, Nucl. Instrum. Methods A **560** (2006) 405.
[8] F. J. Iguaz *et al.*, JINST **6** P07002 (2011).
[9] J.P. Richer *et al.*, Nucl. Instrum. Methods A **620** (2010) 470.
[10] O. Bourrion *et al.*,Nucl. Instrum. and Meth. A**662**(2010) 207.
[11] O. Guillaudin *et al.*, EAS Publ. Ser. **53** 119 (2012).
[12] D. Santos *et al.*, arXiv:0810.1137.
[13] J.F. Ziegler *et al.*, Pergamon Press New York, www.srim.org (1985).
[14] S. Burgos *et al.*, Nucl. Instrum. Methods A **584** (2008) 114.
[15] Q. Riffard *et al.*, Proc. of 48th Recontres de Moriond, VHEPU(2013), arXiv:1306.4173.

NEWAGE

K. Nakamura[1], K. Miuchi[2], T. Tanimori[1], H. Kubo[1], T. Mizumoto[1], J. D. Parker[1], A. Takada[3], H. Nishimura[1], H. Sekiya[4], A. Takeda[4], T. Sawano[1], Y. Matsuoka[1], S. Komura[1], Y. Yamaguchi[2], S. Nakaura[2]

[1] Department of Physics, Kyoto university, Oiwakecho, Sakyo-ku Kyoto-shi, Kyoto, 606-8502, Japan

[2] Department of Physics, Kobe university, Rokodai, Nada-ku Kobe-shi, Hyogo, 657-8501, Japan

[3] RISH, Kyoto university, Oiwakecho, Sakyo-ku Kyoto-shi, Kyoto, 606-8502, Japan

[4] Kamioka Observatory, ICRR, The University of Tokyo, Gifu, 506-1205 Japan

E-mail: nakamura@cr.scphys.kyoto-u.ac.jp

Abstract. NEWAGE (NEw generation WIMP search with an Advanced Gaseous tracking device Experiment)is one of the direction-sensitive dark matter search experiments with a gaseous timeprojection chamber. In this paper, we will report results of recent underground measurements, performancesof our new detector, improvements of head-tail analysis.

1. Introduction

Directionality was said to provide a very strong evidence of direct dark matter detection[1][2]. Since these ideas were proposed, several experimental and theoretical works on the possibility of detecting this distinct signal of dark matter have been performed[3].

NEWAGE is one of these direction-sensitive dark matter search experiments. In order to have directionality, we use a μ-TPC with low pressure gas. We improved a WIMP-proton cross section of 5400 pb for 150 GeV/c^2 WIMPs by a direction-sensitive methods in 2010[4]. However, we need to improve more than three orders of magnitude to set a competitive limits to other direction-insensitive searches. We plan to improve the sensitivity by reducing the internal radioactive background, lowering the energy threshold, discriminating head-tails of nuclear tracks and building large-volume detectors. In this paper, we report these efforts after our first underground run and new result using our latest detector.

2. NEWAGE detectors

We have three TPCs (time projection chamber) for several purpose(table 1). We have two large size detectors, named NEWAGE-0.3a' and NEWAGE-0.3b', in the underground laboratory. NEWAGE-0.3a' took the data for dark matter search in 2012 and improved dark matter limits. A large detector, NEWAGE-0.3b', was installed at Kamioka underground laboratory in March 2012, and the commissioning run is almost done. We have small size detector, NEWAGE-0.1a, in the surface laboratory for advanced test like head-tail R&D.

We use μ-PIC[5, 6], one of the micro patterned gaseous detectors, as a main-multiplier and a readout of the μ-TPC. The pitch of the μ-PIC is 400 μm. We also use a gas electron multiplier

name	μ-PIC	drift	location	use
NEWAGE-0.3a'	$30\times30\,\mathrm{cm}^2$	30 cm	underground	DM run, BG study
NEWAGE-0.3b'	$30\times30\,\mathrm{cm}^2$	40 cm	underground	commissioning
NEWAGE-0.1a	$10\times10\,\mathrm{cm}^2$	10 cm	surface	Head-tail R&D

Table 1. Specifications of NEWAGE TPCs.

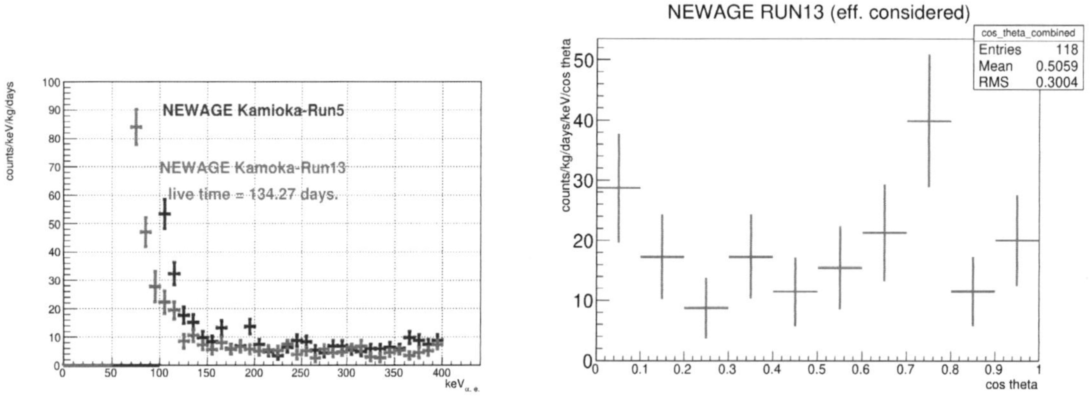

Figure 1. Obtained energy spectrum(left) and the *cosθ* distribution (right) in RUN-13. We unfolded the energy spectrum with detection efficiency and thus this spectrum can be referred as an "effective" energy spectrum. The *cosθ* distribution was made with 100-120 keV events.

(GEM)[7] as a sub-multiplier. For details on our μ-TPC system, please refer to our previous publication[4] and the references therein.

3. Underground measurments (NEWAGE-0.3a')

We performed underground measurements in 2012 at Kamioka underground laboratory with NEWAGE-0.3a. For this dark matter run, we had mainly three detector updates. First, we attached a mini-chamber containing about 100 gram of charcoal (TSURUMICOAL 2GS) at room temperature and circulated the TPC gas to reduce radon background. Second, we replaced glass-reignforced fluoroplastic used for the TPC cage with PTFE to reduce alpha-particles background. Third, we made a more precise gain map to reduce gamma-ray background. Details are described in ref [8].

This measurement, NEWAGE Kamioka RUN-13, started at Jan. 23, 2012 and continued until Jan. 24, 2013. The NEWAGE-0.3a' detector was filled with CF_4 at 152 Torr, and we had 11.48 g in the detection volume. The data acquisition live time was 217.1 days. During RUN-13, we changed the gas six times; once the gas was filled it was continuously circulated two months on average.

We set a fiducial volume of $5\times5\times31\mathrm{cm}^3$ to veto the background from the TPC wall. We eliminated the data with low gas gains and 134 live-days data remained. The gas gain after the live-time cut was stable in $\pm$ 10%. A total exposure used for the further analysis was 0.313 kg days.

We show the energy spectrum and the *cosθ* distribution obtained in RUN-13 in Figure 1. The count rate at 100 $\mathrm{keV}_{\alpha.e.}$ was reduced by a factor of two. From obtained *cosθ* distributions, we set a spin-dependent proton-WIMP cross section of 3130 pb for 150GeV WIMPs.

Figure 2. Photograph of NEWAGE-0.3b' detector.

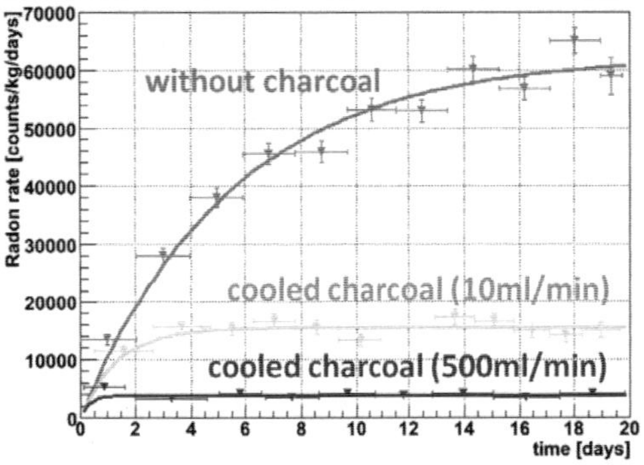

Figure 3. Time dependence of the high energy background rate($3 - 10\,\mathrm{MeV}$). Topmost, middle, bottommost lines are without circulation, 10 ml/min circulation, 500 ml/min circulation respectively.

4. Detector upgrades (NEWAGE-0.3b')

NEWAGE-0.3b' is a new detector designed to improve sensitivities. We made three improvements for NEWAGE-0.3b' construction. One is "large size", another is "radon elimination" and the third one is "low threshold". After the construction and performance test in the surface laboratory, we transfered NEWAGE-0.3b' into Kamioka mine and started taking data. We explain each improvement and commissioning status.

The first improvement is "large size". Not being angular resolution worse significantry, we expanded the drift length from 30 cm to 40 cm. Also, we decided to adopt a larger size GEM to cover whole area of μ-PIC. Due to these improvements, the sensitive volume of NEWAGE-0.3b' is about twice larger than that of NEWAGE-0.3a'. Futhermore, drift cage is made of PEEK, emits less radon gas than our previous material, glass-reignforced plastic.

The second improvement is "radon reduction". Almost all metal slightly contain U-chain or Th-chain isotopes. Since radon background from these isotopes reach secular radioactive equilibrium, radon emanates with a time constant of radon after gas change. Gas circulation is expected to reduce U-chain radon (^{222}Rn) because time constant of such radon is slow as 5.5 days. We constructed a gas circulation system with cooled charcoal in order to reduce U-chain radon. We use a dry-pump to circulate gas and the gas flow is 500 ml/min. We use a refrigerator to cool charcoal, and a heater to control temperature. The temperature in stable operation is 230K. In order to check radon reduction, we measured time depencdence of high energy background. Figure 3 shows time dependence of high energy event rate ($3-10\,\mathrm{MeV}$). Seeing time dependence without gas circulation, we can see the 5 days time constant which indicates existence of U-chain radon in NEWAGE-0.3b detector. Using our gas circulation system, we succeed to reduce radon by 1/10.

The third improvement is "low threshold". In our case, the energy threshold is limited by the track length, because we need a certain length to keep the direction sensitivity. By using the lower pressure gas, we were able to measure the angular resolution in the $50 - 100\,\mathrm{keV}$ energy range in addition to other energy ranges. Thus we lowered the energy threshold to 50 keV keeping direction sensitivity[9].

In March 2013, we transferred NEWAGE-0.3b' to underground laboratory and started comissioning run. First, we measured fundamental detector performances (calibration, drift velocity, flatness), and obtained consistent results compared to the surface measurements. Then, we measured the high energy background without the gas circulation system (Figure

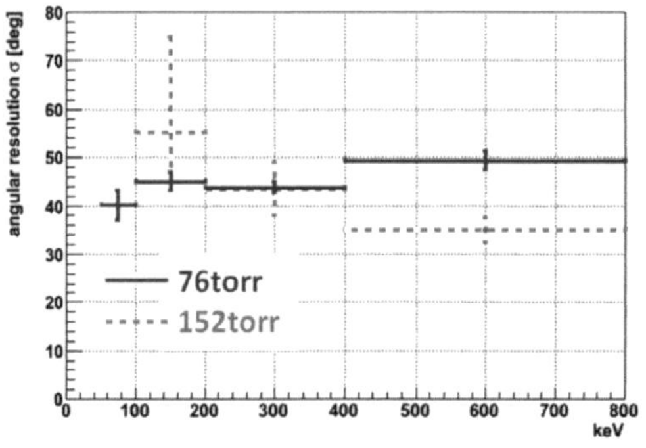

Figure 4. Measured angular resolutions of μ-TPCs. Solid (blue online) line and dashed (red online) line represent result using 76 torr pressurevgas and 152 torr of that respectively.

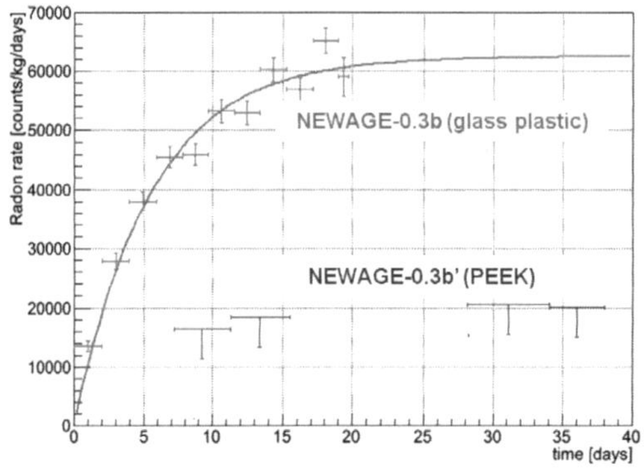

Figure 5. Time dependence of the high energy event rate($3 - 10\,\mathrm{MeV}$). Top line is measured by old drift cage made by glass-reinforced fluoro-plastic (NEWAGE-0.3b) at surface laboratory. Bottom line is measured by new drift cage made by PEEK (NEWAGE-0.3b') at underground laboratory.

5). NEWAGE-0.3b' obtained less than 1/4 lower event rate. By replacing material of drift cage from glass-reinforced fluoro-plastic to PEEK, we reduced radon emanation. After these commissioning run, we started DM run from 17 July 2013.

5. R&D for head-tail recognition (NEWAGE-0.1a)

Head-tail recognition of the nuclear track is important to improve the sensitivity of a direction-sensitive dark matter search experiment[10]. DM-TPC group has shown the possibility of head-tail recognition of high energy nuclear tracks[11] followed by the DRIFT group's work in the energy range relevant to the dark matter search for one dimension[12]. We started the head-tail study just before the last CYGNUS workshop[8]. We used NEWAGE-0.1a with CF_4 gas at 152 Torr. We doubled the statistics and obtained better significances.

When we calculated the skewnesses of the direction parallel to the incoming neutron irradiation, statically significant γ with 4.1σ, 5.1σ, and 7.1 σ are observed for the energy range of 70-100keV, 100-200keV, and 200-400keV, respectively.

When we calculated the skewnesses of the direction orthogonal to the incoming neutron irradiation, the γ were consistent with zero in the all energy ranges.

These results shows that we can recognize head-tail with a sufficient statistics down to 70

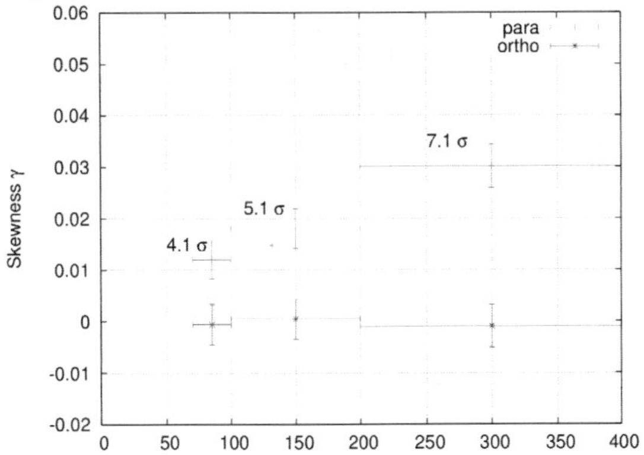

Figure 6. Energy dependence of the measured skewness. The points with errorbars (red online) and asterisks (blue) show the skewness parallel and orthogonal to the incoming neutrons.

keV. We plan to measure the skewness with lower pressure gases to lower the threshold. We also plan to evaluate the skewness from the waveform which contains timing information so that a real three-dimensional head-tail analysis can be carried out.

6. Conclusions

We are proceeding NEWAGE project toward the dark matter detection. From recent measurement taken by NEWAGE-0.3a', we updated WIMP-proton spin-dependent cross setion limit and obtained value was 3130 pb for 150 GeV/c^2 WIMP mass. Aiming to DAMA region as our next step, NEWAGE-0.3b' was constructed, which is low background, low threshold and large volume detector. Now, a commissioning run started at underground laboratory. In future, head-tail recognition is necessary to improve the sensitivity. So we started head-tail analysis with NEWAGE-0.1a and recognized head-tail statistically as for over 70keV.

Acknowledgments

This work was partially supported by KAKENHI (19684005, 23684014, 21340063, and 23 · 812).

References

[1] G. Masek et al. *Proceedings of the Workshop on Particle Astrophysics* (1989) 43.
[2] G. Gerbier et al. *Nuclear Phys. B Proc. Sup.* **13** (1990) 207.
[3] S. Ahlen *et. al. Int. J. Mod. Phys. A* **25** (2010) 1. and references therein.
[4] K. Miuchi *et. al. Phys. Lett. B* **686** (2010) 11.
[5] A. Ochi *et. al. Nucl. Instrm. Methods Phys. Res. Sect. A* **513** (2003) 277.
[6] A. Takada *et. al. Nucl. Instrm. Methods Phys. Res. Sect. A* **573** (2007) 195.
[7] F. Sauli and A. Sharma *Annu. Rev. Nucl. Part. Sci* **49** (1999) 341.
[8] NEWAGE *EAS publication series, Proceedings of CYGNUS 2011, Third International Conference on Directional Detection of Dark Matter* **53** (2012) 33.
[9] K. Nakamura *et. al. JINST* **7** (2012) C2023.
[10] A. M. Green and B. Morgan *Astropart. Phys.* **27** (2007) 142.
[11] D. Dujmic et al. *Nucl. Instrm. Methods Phys. Res. Sect. A* **584** (2008) 327.
[12] S. Burgos and others *Astropart. Phys.* **261** (2009) 266.

Journal of Physics: Conference Series **469** (2013) 012004 doi:10.1088/1742-6596/469/1/012004

Status and analysis system of directional dark matter search with nuclear emulsion.

T.Katsuragawa*, T.Naka[a,b], T.Asada, M.Yoshimoto, K.Hakamata, M.Ishikawa

Department of Physics, Nagoya University, Nagoya 464-8602, Japan
[a]Institute for Advanced Research, Nagoya University, Japan
[b]Kobayashi-Maskawa Institute for The Origin of Particles and The Universe, Nagoya University, Japan

E-mail: katsuragawa@flab.phys.nagoya-u.ac.jp

Abstract. We have been doing research and development for direct dark matter search by nuclear emulsion which is a solid state detector. This experiment enable directional detection of dark matter with the large mass target and model independent. Until now, we constructed a base of fully automatic analysis system and nuclear emulsion which can detect sub-micron tracks. We have demonstrated that it is possible to detect recoiled tracks of 100 nm or more by neutron irradiation. This track length is correspond to 37 keV in C(N,O) target. Additionally, we evaluated the angular resolution of the energy basis by using an ion implant system, and obtained 25 degrees or better resolution in 80 keV carbon ions. The fully automatic analysis system which can analyze very short tracks lead the experiment to next phase, we will do a quantitative study of the background toward gram scale test experiment at the Gran Sasso underground laboratory.

1. Introduction

We are aiming to detect dark matter around galaxy with the nuclear emulsion detector (emulsion). Then, dark matter signals are nuclei which is recoiled by dark matter in the emulsion, and the emulsion can detect them together with directional information. Because, emulsion has extremely high spatial resolution and solid-state detector (3.2 g/cm^3). So, this experiment enable directional dark matter search with the large mass target and model independent. On the other hand, the emulsion don't have time resolution. So, we will control the direction of the emulsion to Cygnus by mounting on the equatorial telescope. First, we aim to conclude about parameter region (dark matter-nucleon cross section of $10^{-40}\sim 10^{-41}$cm^2 and dark matter mass of 10 GeV/c^2) with direction information which is claimed dark matter signals by preceding experiments (DAMA [1], CoGeNT [2] and CRESST [3]).

2. Detector

The emulsion is a kind of photographic film which made from silver halide crystals and polymer as a binder (Table 1). When a charged particle penetrates some crystals, track is detected as a line of silver grains by a process of chemical development. Now, emulsion is able to be produced in Nagoya university Japan and, various types of emulsions can be produced and studied by this facility. In the case of dark matter search, it is necessary to detect the sub-micron scale tracks. Then, small crystal size is very essential to realize lower threshold of detectable range (energy). New type emulsion detector with fine-grained crystals (40 nm) was already confirmed to be able to detect tracks of more than 100 nm (Figure 2) by using low velocity ions due to ion implantation system like a recoiled nuclei (Figure 1).

Table 1. Parameter of the emulsion.

Atom	Mass ratio [%]
Ag	46
Br	34
C (N, O)	19
Total mass density	3.2 g/cm^3

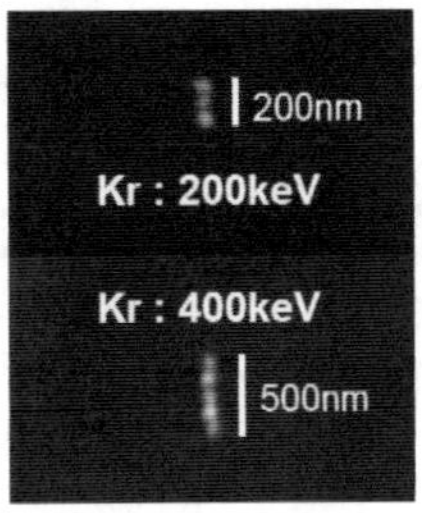

Figure 1. Sub-micron track images by using Kr ion with electron microscope.

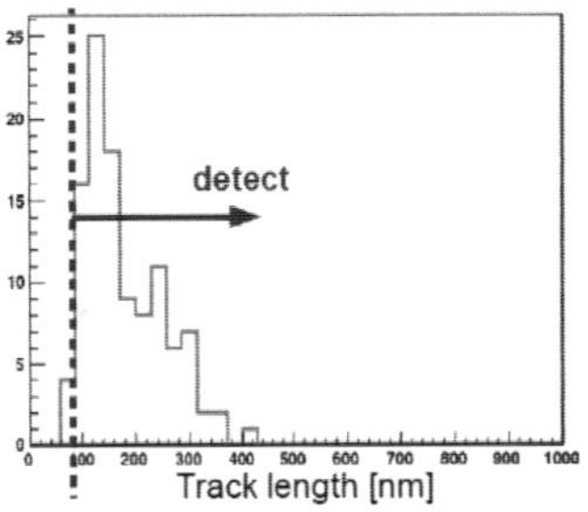

Figure 2. Detectable track lengths are 100 nm or longer by measurements in the electron microscope.

3. Readout system

Analysis is divided into two main phases. In the first scan, we use the optical microscope for selection of candidate. Because, the optical microscope can analyze large volume. In the second scan, candidates are confirmed whether a track or not by using higher spatial resolution microscopy (e.g. X-ray microscope, electron microscope, and super-resolution technique with optical microscope).

3.1. Readout technique of sub-micron tracks

First, the development of readout system to parameterize the quality of track as an Ellipticity (Major/Minor) have been done by constructing the prototype optical microscope called "PTS" for R&D. By this shape recognition system, sub-micron tracks with two or more silver grains can be selected in the emulsion, because main noises are one silver grain and looks like sphere with optical microscope.

The performance of the emulsion sample exposed to neutrons (14.8 MeV by D-T reaction) was evaluated, because recoiled nuclei tracks due to neutron are similar to that of scattered by dark matter. By using the PTS, we have examined the ellipticity distribution of events which have already been confirmed as tracks or noises by X-ray microscopy analysis collaborated with Spring-8 in Japan [4]. So, we constructed X-ray microscope for second scan, and which is good tool to confirm the candidates. Because, X-ray microscope has higher spatial resolution and non destructive observation is possible. And, we could match the events between optical and X-ray microscope like Figure 3. Then, matching efficiency was achieved more than 99 %, and the plot of major vs. minor was obtained like Figure 4 which show signal region by the ellipticity. As the result, the PTS has about 70 % readout efficiency by the method of ellipse fit about 200 nm tracks compared to the X-ray microscope, and 320 nm or longer tracks are 100 % that was shown in Figure 5. And, the angular distribution of tracks in signal region indicated the direction of neutron (Figure 6).

In addition, the best focus selection system for the shape recognition was also developed. So, a base of analysis system was complete. Emulsion can be analyzed fully automatic with optical microscope. Here, current maximum readout speed is about 1 g/10day.

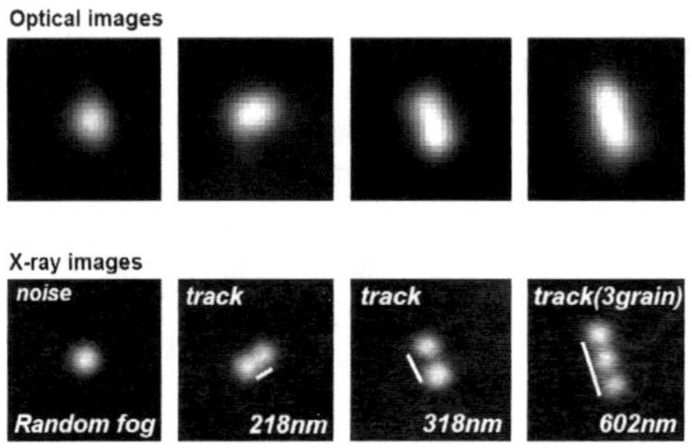

Figure 3. The pinpoint checking of events and noise between optical and X-ray microscope. Above: optical microscope images. Below: X-ray microscope images.

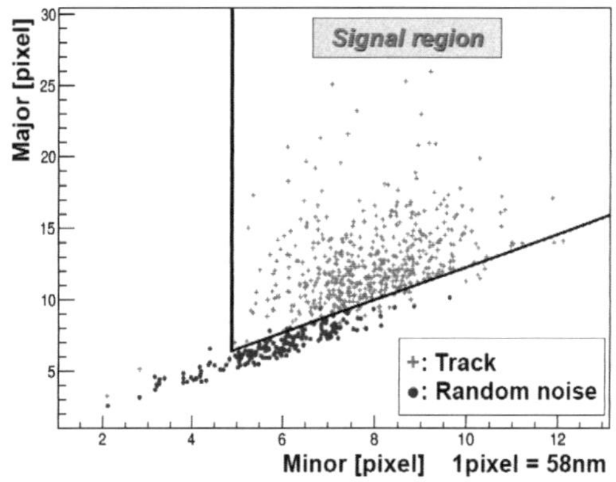

Figure 4. The data is optical readout result of events already confirmed as tracks and noises by X-ray microscope. Signal region that is determined by the Ellipticity (Major/Minor). + dots (red dots in online version) are result of tracks due to neutrons, and • dots (blue dots in online version) are noise events.

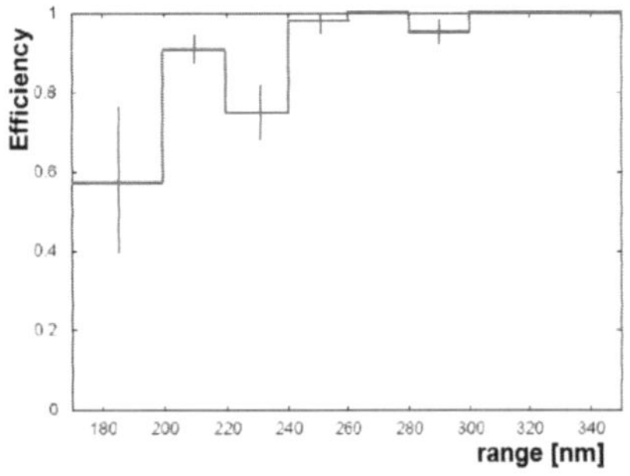

Figure 5. Readout efficiency with optical microscope compared to X-ray microscope. This data was obtained by result of Figure 4, and still preliminary.

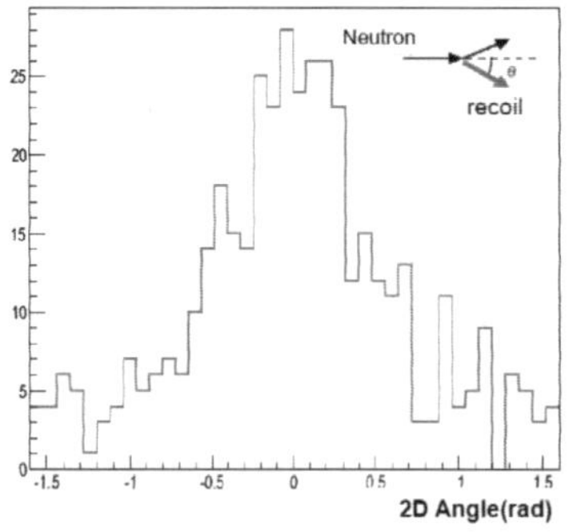

Figure 6. Angular distribution of recoiled tracks in signal region by image processing with optical microscope. This result is consistent with the incident direction of neutrons.

3.2. Angular resolution with optical microscopy

We confirmed angular resolution of optical readout by using ion-implantation system. Ion-implantation system can be used as a demonstration of monochromatic energy nuclei recoiled. In this time, carbon ions are used, and the angular resolution was obtained by compared to the Gaussian distribution of the Monte Carlo simulation (SRIM) [5] smeared by addition of the angular resolution from ideal incident angular distribution of the carbon ions. Angular resolution with optical microscope is better than about 25 degree for 80 keV carbon recoiled tracks like Figure 7.

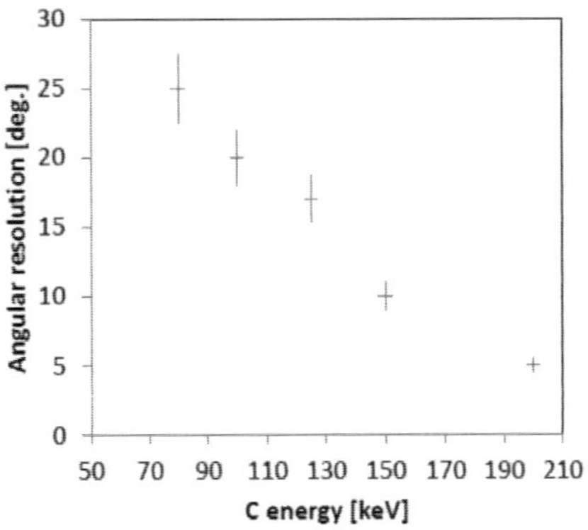

Figure 7. Angular resolution of carbon ions due to ion implantation system, and energy of carbon ions can be selected. This data is still preliminary, because there is a possibility to change depending on improvement of the emulsion and chemical development.

3.3. Techniques to improve the readout efficiency

We have expansion technique to improve the energy threshold with optical microscope [6]. By expansion of emulsion, tracks are elongated. It is possible to expand up to two times by chemical treatment. Therefore, more than 100 nm tracks can be selected. Then, distortion of angle is 0.41 degree after expansion and this is small enough compared to angular resolution.

4. Summary and near future plan

Emulsion have been developed for dark matter search which can detect the tracks of 100 nm or more, and a base of fully automatic analysis system was also completed. The tracks of 100 nm are correspond to about 37 keV C(NO) recoils. As the next step, we will start to study about background (β-rays, γ-rays, and neutrons), and preparing of the measurement about radio activity in the our material was already started. In addition, the rejection efficiency about background with current emulsion will be obtained by the PTS, and we will do R&D of the emulsion about background and low velocity ions sensitivity. In parallel, we are aiming to improve the readout efficiency by changing the conditions of the microscope in order to construct the next optical microscope with higher spatial resolution and readout speed, and to start first test experiment with gram scale emulsion in Gran Sasso national laboratory (LNGS, Italy).

References

[1] R.Bernabei et al, Eur. Phys. J.,C67, 39 (2010)
[2] C.E. Aalseth et al., Phys.Rev.Lett, 107, 141301 (2011)
[3] G.Angloher et al., Eur. Phys. J C, 72, 1971(2012)
[4] Y.Suzuki et al., Jpn. J. Appl. Phys. 44(2005) 1994-1998.
[5] J.F.Ziegler and P.Biersack, Pergamon Press, NewYork (1985)

[6] K.Kimura and T.Naka NIM A 680 (2012) 12-17
[7] T.Naka et al, NIM A 718 (2013) 519-521

MPGD R&D Activities in JAPAN

Atsuhiko Ochi
Kobe University, Kobe 657-8501, JAPAN
E-mail: ochi@kobe-u.ac.jp

Abstract.
Micro Pattern Gaseous Detector (MPGD) is one of key technology for fine particle tracking and imaging. Some part of recent MPGD R & D activities in Japan are reviewed.

1. Introduction

Micro Pattern Gaseous Detectors (MPGDs) are a relatively new kind of particle detector, which are based on gaseous multiplication using micro pattern electrodes instead of wires in wire chambers. The first MPGD; Micro Strip Gas Counter (MSGC), was invented by Anton Oed in 1988 [1], after that many type of MPGDs have been proposed. At the end of the 1990's, technical break-throughs were brought by invention of the GEM and the Micromegas [2, 3]. Those new two types of detectors were used in COMPASS as the first MPGDs in high energy physics experiments [4, 5], which achieved great success. Nowadays, more various types of MPGDs (e.g. MHSP, μ-PIC, THGEM,) are being actively developed [6, 7, 8]. These technologies allow for detectors with unprecedented spatial resolution, high-rate capability, large sensitive area, operational stability and increased radiation hardness. A large number of groups worldwide are developing MPGD devices for future experiments at particle accelerators, for experiments in nuclear and astroparticle physics, as well as for industrial applications such as medical imaging, material science, and security inspection.

The MPGD developments in Japan is very active as well as in Europe. From the early 1990's, the MSGC and capillary plate was developed for X-ray imaging and particle tracking [9, 10]. The μ-PIC is originally invented in Japan at 2001[7], and it is using for many type radiation imaging detector now. Recently, many scientific groups develop and/or apply the MPGD for particle/radiation detector in Japan. More than 70 participants are joined to annual Japanese MPGD workshop, which has been held since 2004. In this article, the activities related on MPGDs are reviewed as it was presented in CYGNUS2013.

2. MPGD applications

There are many aspects on MPGD applications in Japan. Using the high rate capability, time resolved X-ray imaging were realized by Kyoto Univ. using μ-PIC and GEM as shown in figure 1(a). The figure 1(b) shows small angle X-ray diffraction pattern. using this detector. Very wide ($\sim 10^5$) dynamic range can be displayed by using high speed data acquisition system [11].

Time resolved imaging is also very useful for neutron imaging. KEK group has been developing time resolved neutron radiography systems using GEM with Boron (figure 2(a)). The neutron spectroscopic imaging are available in spallation neutron source, for separating timing

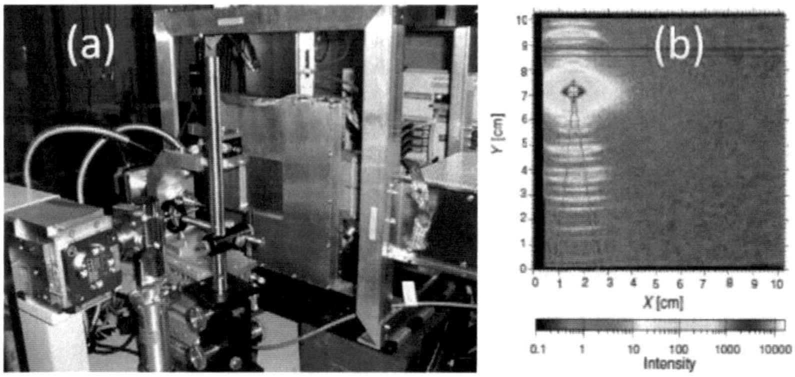

Figure 1. Picture of the time resolved X-ray imaging detector using μ-PIC (a), and the X-ray small angle diffraction image using synchrotron radiation (b). [11]

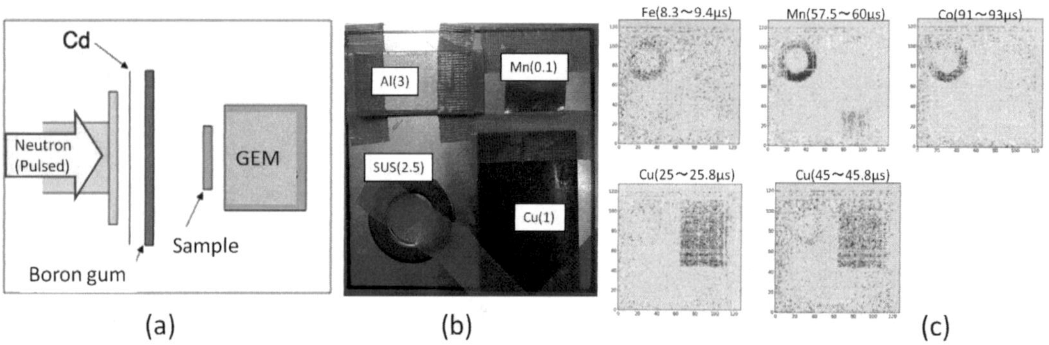

Figure 2. Neutron radiography using boron GEM at KEK (a). The sample with different metal on one sheet (b) is set in front of the GEM. The neutron pulse from spallation source is arrived at sample with time distribution, which correspond to neutron energy. Differences of neutron absorption spectrum for each elements are observed time resolved neutron imaging as shown (c).

window, that correspond the neutron time of flight. Figure 2(c) shows the neutron transparent image of sample (figure 2(b)). The different images of each time window correspond to different neutron energy [12].

The MPGD properties of both fine position and timing resolutions make possible the small time projection chambers (TPCs), by which sub-millimeter size of charged particle are traced. Using the "fine" TPCs, gamma ray camera, fine neutron imaging and dark matter wind detector are developed using combination of GEM and μ-PIC. The gamma-ray camera is realized by ETCC (Electron Tracking Compton Camera) method. Figure 3 shows the schematic principle of operation of ETCC. The arrival directions of incident gamma-ray (from several hundred keV to several MeV) can be resolved by measuring both Compton scattered gamma-ray (energy and position) by scintillator and track of recoiled electron by small TPC. Those cameras are developing for the field of gamma astronomy and medical imaging in Kyoto University. [13, 14]. Fine neutron imaging is realized by observing proton and triton tracking which is produced by n(^{3}He, T)p reaction in small TPC. Figure 4 shows transparent image of the wrist watch [15]. For dark matter search, it is thought that trajectories of recoiled nucleons by WIMP can be observed using small TPC. Details of this type experiments will be described in other articles in this proceedings.

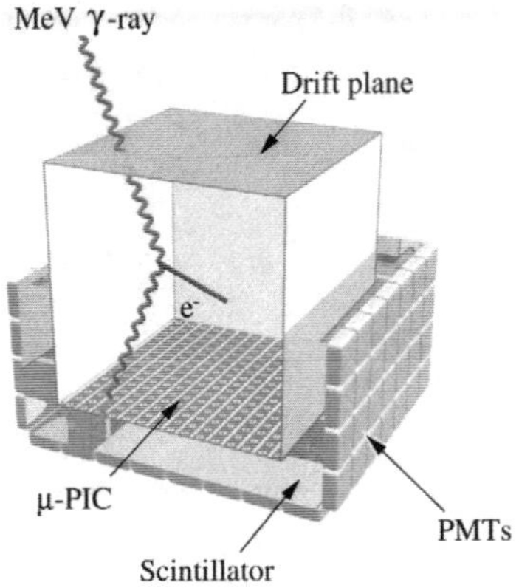

Figure 3. Schematic principle of operation of ETCC method using μ-PIC and scintillators. Both scattered gamma and recoiled electron track are measured at one event. Then, incident direction of gamma-ray can be resolved.

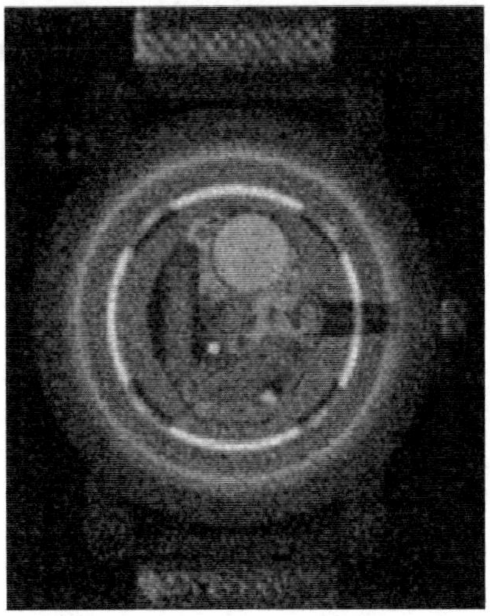

Figure 4. Fine neutron imaging demonstrated by transparent imaging of the wrist watch. This picture was obtained by μ-PIC micro TPC with ^{3}He gas.

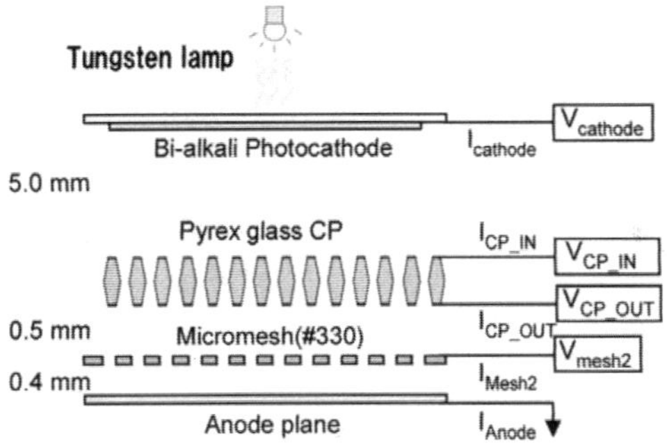

Figure 5. Picture of gaseous PMT and schematic principle of operation, which is developed in Yamagata Univ., TMU and HAMAMATSU. Photo electrons in gas are multiplied by glass capillary and metal mesh.

Not only the particle imager/tracker, photo detection is one of topical application of the MPGDs. The gaseous PMTs using micro mesh and glass capillary plate are developed by Yamagata Univ., TMU and HAMAMATSU group (Figure 5) [16]. Large effective area with moderate position and timing resolution has been achieved, and also it can be easily operated under a very high magnetic field. This type of photodetector is promising for holding good properties of both conventional PMTs and semiconductor detector.

3. Material studies for MPGD

Nevertheless the MPGD has already been used for many applications, many requirements are remained for future experiments. One of most critical problem is spark problem between the electrodes. It is well known that the occasional discharges are invoked when the total electrons (primely electrons times gas multiplication) exceed 10^{7-8} in one event, which is called as "Raether limit" [17, 18, 19]. Those sparks can potentially damage the electrodes. A number of sparks (for some situations, only one spark) makes the conductive path between HV electrodes

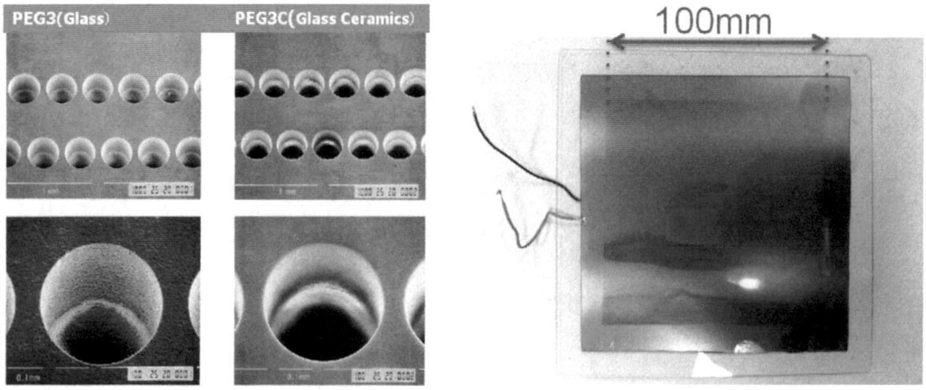

Figure 6. Microscopic picture of small holes and over view of glass GEM. Those are fabricated by using PEG3 glass provided by HOYA corp.

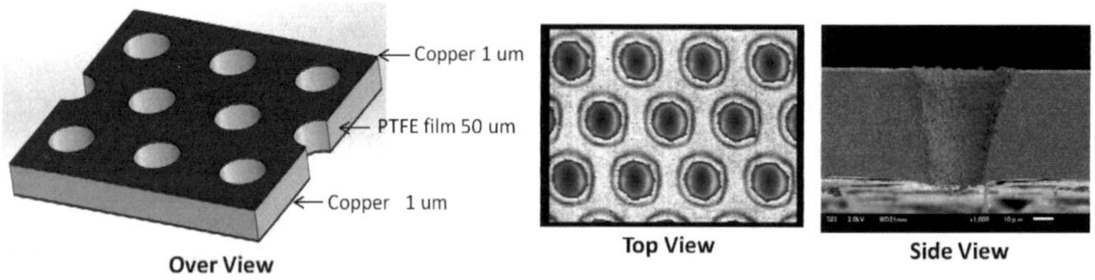

Figure 7. Schematic view and microscopic views of PTFE GEM. Substrate is made from PTFE with 50 μm thickness, and the holes are drilled by laser.

on the substrate, and it cause permanent damage to the detector.

There are mainly two way to solve the discharge problem. One is to make the MPGD substrate for holding tolerances against the sparks. Another one is to make the MPGD electrodes for reducing damage from sparks. Those researches and developments are under intense investigation by some of Japanese MPGD group with unique method.

For the substrate of the MPGDs, the polyimide has been used conventionally in many cases because of the tolerance to the high voltage, radiation and heat. Also machining performances are rather good for making micro pattern. However, since polyimide is made from organic material, it can be carbonized by sparks on the surface. Therefore, there are researches for using substrate with non carbonized material.

Univ. of Tokyo group has made the GEM substrate by special glass. Their material is photo-etchable, called PEG3, made by HOYA cooperation. Figure 6 shows the microscopic picture of small holes and over view of glass GEM. They have made GEM (THGEM) with thickness of 680 μm and hole diameter of 170 μm using PEG3. The maximum gas gain is attained to close to 10^5 under Ne/CF_4 (90:10) mixed gas. This glass is not only non-organic material, but also outgas free, then it is expected to use for sealed gas application.

Tokyo IRI and RIKEN group has developed GEM with PTFE (polytetrafluoroethylene) substrate. They made the holes by laser drilling, and form the electrodes by copper sputtering. Thickness of the substrate is 50μm, which is same as standard (conventional) GEM. Figure 7 shows schematic view and microscopic views of their PTFE GEM. Their preliminary tests show the gas gain of more than 2×10^4, which is about 10 times grater than conventional GEM with same substrate thickness. The most characteristic property of this GEM is the tolerance to the sparks. No damages were observed by more than 20000 sparks in 20mm $\times$ 20mm prototype.

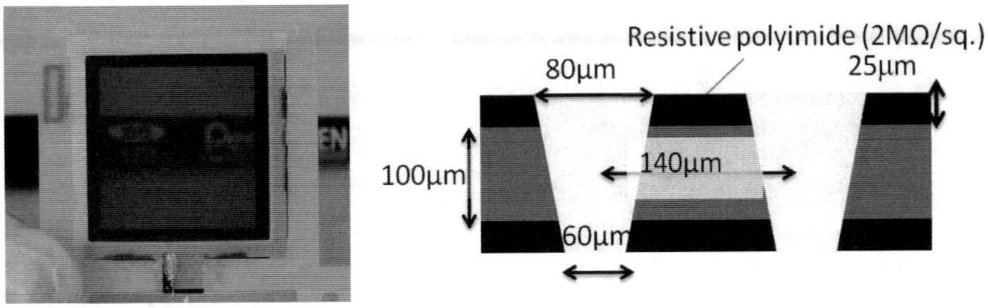

Figure 8. Overview of resistive GEM and its schematic structure.

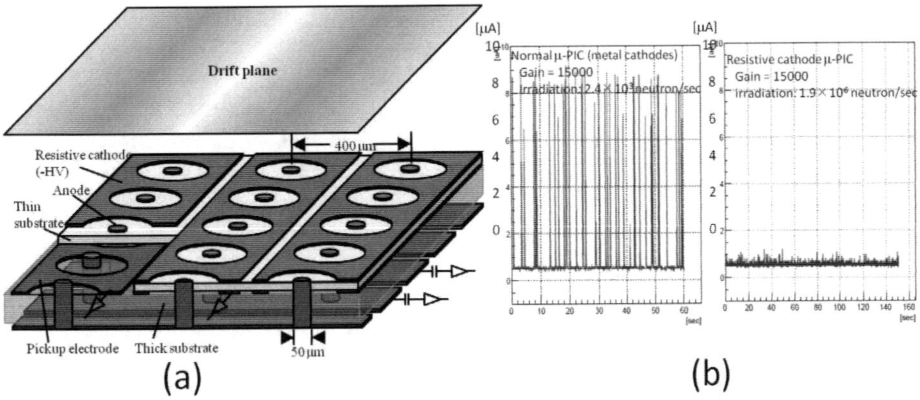

Figure 9. Schematic structure of resistive μ-PIC (a), and spark monitor under intense neutron for both resistive and conventional μ-PIC. Note that the neutron intensity in resistive one is about 10^3 times greater than conventional one.

4. Resistive electrodes

The damages by sparks are not only the detector structure, but also readout electronics and operation condition while HV recovery. It cause long dead time for detection. To reduce the influence of the spark, there are researches to make the electrodes with resistive material.

RIKEN group has developed GEM foil with resistive electrodes, those are replaced from conductive layers[20]. Figure 8 shows picture of resistive GEM and schematic structure. The resistive electrodes are made from carbon loaded polyimide foils, and those are attached to LCP foils of 100μm thickness on double side. The holes are drilled by laser with conical shape of 60-80 μm diameter. It was the first resistive GEM with attaining high (> 300) gas gain.

The μ-PIC with resistive electrodes has been developed at Kobe University group [21]. As shown in figure 9(a), cathode electrodes are replaced with resistive material, which is made from carbon loaded polyimide. The signals are read both from anodes and pickup electrodes, those are lied under cathode via insulator. The spark between anode and cathode is strongly suppressed due to voltage drop on resistive cathode. Figure 9(b) shows anode HV current monitor under intense fast neutron (up to $\sim 10^6$ neutron / second) of resistive μ-PIC and conventional μ-PIC. Spikes in this graph mean the sparks. From these test results, the spark rate of resistive μ-PIC was reduced 10^{3-5} times comparing with using conventional μ-PIC.

The MicroMEGAS with resistive anodes are developed for LHC upgrade at Kobe Univ. and ICEPP. The signals are read as a induced charge by readout strips those are printed under substrate. Figure 10 shows the schematic structure of MicroMEGAS with sputtered electrodes, and the figure of detector (without gaseous vessel). It should detect the MIP (muon) under intense HIP particles. Strong reduction of the spark probability is essential, and this type of

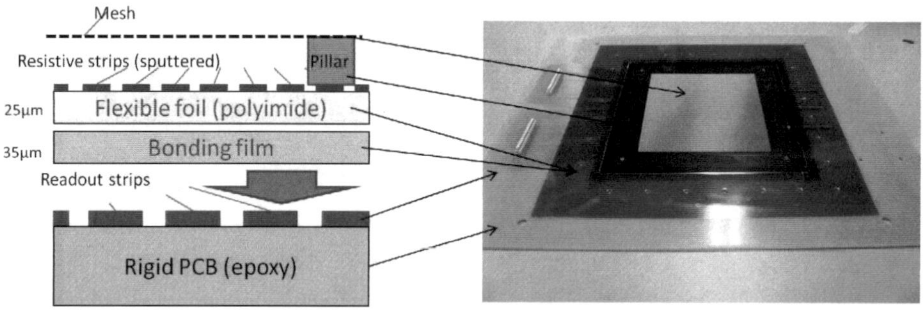

Figure 10. Schematic structure and picture of resistive MicroMEGAS, which is developed as a prototype of ATLAS upgrade. Resistive layer is formed by carbon sputtering on the polyimide.

resistive MicroMEGAS have shown very promised results in many tests. For using this as a part of ATLAS detector, it is also necessary to establish the way for making large size with low cost. Kobe and ICEPP have developed two new production methods for resistive strips. One is screen printing, and another is carbon sputtering. Screen printing makes rather rough patterns, however the very low cost is available in production. On the other hand, sputtering with lift off method makes very fine patterns, and resistivity control is also accurate (within 10% at 500kΩ/sq.). The production cost is moderate, but it is higher than the screen printing.

5. Summary

The MPGD researches and developments are growing up in Japan. Both application developments and basic detector studies are very active. The variety type of MPGDs are developed in various field of science and technology with their original ideas. Especially, the material studies of substrates and electrodes are growing recently, and those studies will be expected to open the new field of experimental physics. I would also like to remark that the introduced works are only some part of Japanese MPGD activities. There are many other attractive MPGD studies in Japan.

References

[1] A. Oed, *Nucl. Instrm. Meth.***A 263** (1988) 351
[2] I. Giomataris, et al., *Nucl. Instrm. Meth.***A 376** (1996) 29
[3] F. Sauli, et al., *Nucl. Instrm. Meth.***A 386** (1997) 531
[4] B. Katzer, *Nucl. Instrm. Meth.***A 494** (2002) 142
[5] P. Abbon, et al., *Nucl. Instrm. Meth.***A 577** (2007) 455
[6] J.F.C.A. Velso, et al., *Rev. Sci. Instrm.***71** (2000) 2371
[7] A. Ochi, et al., *Nucl. Instrm. Meth.***A 471** (2001) 264
[8] R. Chechik, et al., *Nucl. Instrm. Meth.***A 553** (2005) 35
[9] T. Nagae, et al., *Nucl. Instrm. Meth.***A 323** (1992) 236
[10] H. Sakurai, et al., *Nucl. Instrm. Meth.***A 374** (1996) 341
[11] K. Hattori, et al., *Journal of Synch. Rad.* Vol.16, Part 2 (2009) 231
[12] S. Uno, et al., *Phys. Procedia* **37** (2012) 600
[13] A. Takeda, et al., *Nucl. Instrm. Meth.***A 546** (2005) 258
[14] A. Takeda, et al., *J. of the Phys. Soci. of Japan*, **78**(2009) Suppl. A, pp. 161
[15] T. Tanimori, et al., *Nucl. Instrm. Meth.***A 529** (2004) 373
[16] F. Tokanai, et al., it Nucl. Instrm. Meth.**A 610** (2009) 164
[17] H. Raether, *Electron Avalanches and Breakdown in Gases.* London, UK: Butterworth, 1964.
[18] A. Bressan, et al., *Nucl. Instr. Meth.***A 424** (1999) 321
[19] S. Bachmann, et al., *Nucl. Instr. Meth.***A 479** (2002) 294
[20] A. Yoshikawa, et al., *JINST***7**(2012) C06004
[21] A. Ochi, et al., *JINST***7**(2012) C05005

Measurement of the electron drift velocity for directional dark matter detectors

F. Mayet[1], J. Billard[2,3], G. Bosson[1], O. Bourrion[1], O. Guillaudin[1], J. Lamblin[1], J. P. Richer[1], Q. Riffard[1], D. Santos[1], F. J. Iguaz[4], L. Lebreton[5], D. Maire[5]

[1]Laboratoire de Physique Subatomique et de Cosmologie, Université Joseph Fourier Grenoble 1, CNRS/IN2P3, Institut Polytechnique de Grenoble, 53, rue des Martyrs, Grenoble, France
[2] Department of Physics, Massachusetts Institute of Technology, Cambridge, MA 02139, USA
[3] MIT Kavli Institute for Astrophysics and Space Research, Massachusetts Institute of Technology, Cambridge, MA 02139, USA
[4] Univ. de Zaragoza - Dep. de Fisica Teorica, Pedro Cerbuna 12, E-50009 Zaragoza, Spain
[5] LMDN, IRSN Cadarache, 13115 Saint-Paul-Lez-Durance, France

E-mail: mayet@lpsc.in2p3.fr

Abstract. Three-dimensional track reconstruction is a key issue for directional Dark Matter detection. It requires a precise knowledge of the electron drift velocity. Magboltz simulations are known to give a good evaluation of this parameter. However, large TPC operated underground on long time scale may be characterized by an effective electron drift velocity that may differ from the value evaluated by simulation. *In situ* measurement of this key parameter is hence a way to avoid bias in the 3D track reconstruction. We present a dedicated method for the measurement of the electron drift velocity with the MIMAC detector. It is tested on two gas mixtures : CF_4 and CF_4 + CHF_3. We also show that adding CHF_3 allows us to lower the electron drift velocity while keeping almost the same Fluorine content of the gas mixture.

1. Introduction

Directional detection of galactic Dark Matter offers a unique opportunity to identify Weakly Interacting Massive Particle (WIMP) events as such [1–13]. This new search strategy requires the simultaneous measurement of the recoil energy and the direction of the 3D track of low energy recoils. There is a worldwide effort toward the development of a large TPC (Time projection Chamber) devoted to directional detection [14, 15]. All current projects [16–20] face common challenges amongst which 3D track reconstruction [21] is the major one.

It is compulsory to have a precise knowledge of the electron transport properties in the gas mixture used as a sensitive medium for the TPC [22], in particular the electron drift velocity. Indeed, primary electrons, created along the recoil trajectory, are used to retrieve the recoil track in the TPC. In particular, for the MIMAC project [19], the measurement of the third dimension, along the electric field, is achieved thanks to a sampling of the primary electron cloud. Large TPC operated underground on long time scale may be characterized by an effective electron drift velocity that may differ from the value evaluated by Magboltz simulation [23], due to *e.g.* impurities, field inhomogeneities, long drift distances. *In situ* measurement of this key parameter is hence a way to avoid bias in the 3D track reconstruction.

Standard measurements of the electron drift velocity are done thanks to photo-electrons generated by a N_2 laser [24–26], by measuring the time difference between the UV emission time and the electron arrival time on the anode. In order to have a precise measurement, small drift spaces are used [26], $\mathcal{O}(10)$ mm, ensuring that the electric field remains homogeneous. This method is not suited for large drift spaces and *in situ* measurement allows us to retrieve the effective drift velocity that electrons are encountering in the Dark Matter detector.
We present a dedicated method for an *in situ* measurement of electron drift velocity with the MIMAC detector [19], using a collimated α source, together with a maximum likelihood method associated to a modelisation of the signal induced on the grid.

2. Experimental setup

This electron drift velocity measurement is done directly with the MIMAC detector [19]. The primary electron-ion pairs produced by a nuclear recoil in the MIMAC chamber are detected by drifting the primary electrons to the grid of a bulk Micromegas [27] and producing the avalanche in a very thin gap 256 μm. The MIMAC prototype μTPC is composed of a pixelized anode featuring 2 orthogonal series of 256 strips of pixels (X and Y) and a micromesh grid separating the amplification (grid to anode) and the drift space (cathode to grid). Each strip of pixels is monitored by a current preamplifier and the fired pixel coordinate is obtained by using the coincidence between the X and Y strips (the pixel pitch is 424 μm). In order to reconstruct the third dimension of the recoil, along the drift axis, a self-triggered electronics has been developed [28, 29]. It allows us to perform the anode sampling at a frequency of 50 MHz. Hence, the effective electron drift velocity need to be known, which is the goal of this paper.
For the electron drift velocity measurement with the MIMAC detector, we use a collimated α source ($E_\alpha = 5.478$ MeV from ^{241}Am), positioned on the cathode and facing the anode. As the energy loss is about 3 MeV in the TPC (50 mbar of CF_4), the α particle is expected to cross the whole drift space (17.7 cm) as well as the 256 μm amplification space. Since the α particle velocity is much greater than the electron drift velocity by about 2 orders of magnitude, the α arrival time on the anode is taken as the starting time of primary electrons at the cathode. The usual MIMAC amplification field has been lowered by a factor of 1.5, the α ionisation energy is around 3 MeV, much bigger than for low energy nuclear recoil.
In order to recover a non biased estimate of the drift velocity, we have measured the time dependent charge collection profile. The rise time of the charge sensitive preamplifier (when charges are injected on the grid), is about 400 ns. This is much lower than the expected collection time of the primary electrons coming from the alpha tracks for the different gases and drift field considered hereafter, which are estimated using Magboltz to be between 1.5 μs and 17 μs for a pure CF_4 gas and a CF_4 + CHF_3 gas mixture respectively. Hence, we should be able to measure an accurate charge collection profile for each alpha track observed.

3. A likelihood-based data analysis strategy

Straightforward data analysis strategies would not allow us to estimate the electron drift velocity without bias. For instance, one may evaluate the electron drift time, corresponding to a given drift length, with the charge preamplifier, connected to the grid, by measuring the time between the maximum and the minimum. In this case, a significant lengthening of the output signal is expected, due to the readout time constant. This leads to an under-estimation of v_d. The information contained in the 3D track may also be used through the use of the time difference between the first and the last spatial coincidence. While being almost not delayed by the electronic readout, this estimation is expected to depend heavily on the amplification electric field (the gain). Indeed, the probability to have a spatial coincidence depends on the number of electrons contained in a given time sample and hence on the amplification gain.
To avoid bias due to electron diffusion, ion collection time and electronic readouts, we propose

Journal of Physics: Conference Series **469** (2013) 012006 doi:10.1088/1742-6596/469/1/012006

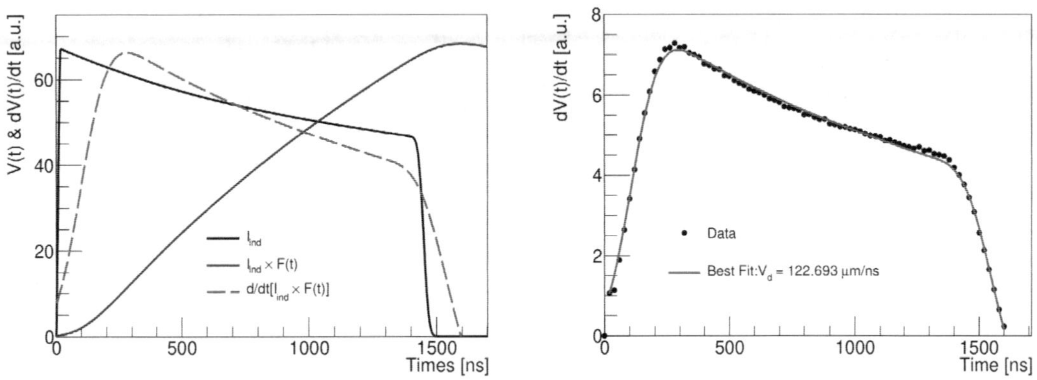

Figure 1. *(Left) Theoretical signal* $V_{\rm th}(t)$ *(blue curve) from the convolution of the current induced on the grid* $I_{\rm ind}(t)$ *(blaks curve) and the transfer function of the charge preamplifier* $F(t)$*. The dashed red curve is the time derivative* $V'_{\rm th}(t)$ *of the signal. (Right) Mean profile (black point) and best fit (red curve) for the time derivate of the signal* $V'(t)$*. The measurement has been done for a pure* CF_4 *gas at* 50 *mbar, with* $E_d = 137.9$ *V/cm and* $E_a = 14.5$ *kV/cm.*

a data analysis strategy based on a profile likelihood method. It requires a complete modeling of the signal, from the current induced on the grid to the measured signal $V(t)$, in order to compare data and expected signals. The theoretical signal $V_{\rm th}(t)$ is given by [30]

$$V_{\rm th}(t) \propto \int\int\int \frac{dE}{dt}(t-\xi) \times Q_{\rm ion}(\xi - \tau) \times g_{\rm diff}(\tau - T) \times F(T)\ dT d\tau d\xi \qquad (1)$$

where :

- dE/dt is the collection of primary electrons on the anode, *i.e.* the time projection, for a given electron drift velocity v_d, of the ionisation energy loss dE/dx simulated with Geant 4 [31].
- $Q_{\rm ion}(t)$ is the ion-induced current, approximated by a gate function of width $\Delta t_{\rm ion} = \epsilon / v_{d_{\rm ion}}$, where $v_{d_{\rm ion}}$ is the drift velocity of ions in the amplification space and ϵ is the Micromegas amplification width (256 μm) [27].
- $g_{\rm diff}(t)$ is a gaussian distribution accounting for the electron longitudinal diffusion, with a standard deviation $\sigma_l(t) = D_l\sqrt{v_d \times t}$, where D_l is the longitudinal diffusion coefficient.
- $F(t)$ is the measured charge sensitive preamplifier transfer function [30]

Figure 1 presents the theoretical signal $V_{\rm th}(t)$ (blue curve) obtained from the convolution of the induced current on the grid $I_{\rm ind}(t)$ (black curve) and the transfer function of the charge preamplifier $F(t)$ [30]. The time derivate $V'_{\rm th}(t)$ of the signal $V_{\rm th}(t)$ is also presented (dashed curve). For this example, the following parameter values are used : $D_l = 440\ \mu{\rm m}/\sqrt{\rm cm}$, $v_{d_{\rm ion}} \sim 8\ \mu$m/ns and $v_d = 122\ \mu$m/ns. In the following, the theoretical signal $V_{\rm th}(t; v_d, v_{\rm ion}, D_l)$ is the adjusting model used in the profile likelihood method that allows us to estimate the uncertainty on the electron drift velocity. This way the uncertainties on all other parameters are accounted for, leading to a the profiled likelihood function that is larger ($\sim$ 40 %) than the standard one. Including systematics from the experimental setup, such as the length of the chamber and the homogeneity of the electric field along the drift space, the uncertainty on the electron drift velocity is evaluated at the level of 1%.

For each setup configuration, we measure $\sim$ 500 α tracks and we evaluate a mean profile $\bar{V}(t)$

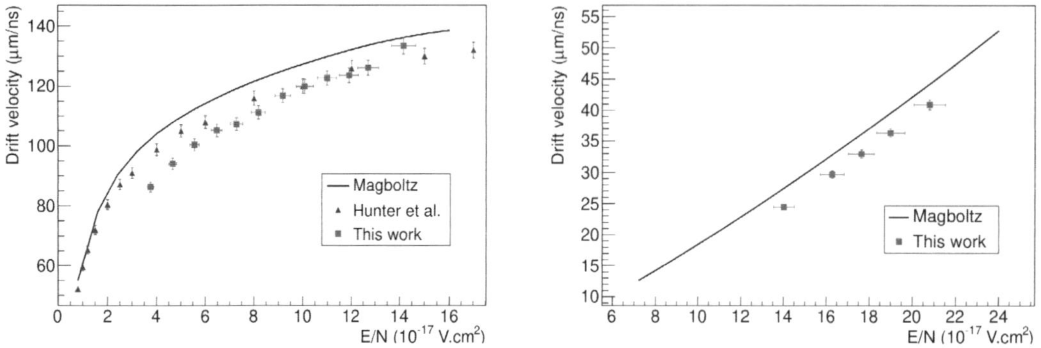

Figure 2. *Electron drift velocity v_d (μm/ns) in CF_4 (left) and in a 70% CF_4 + 30% CHF_3 gas mixture (right) as a function of E/N (10^{-17} Vcm^2), at 50 mbar and for an amplification field $E_a = 15.6$ kV/cm for CF_4 + CHF_3 and $E_a = 14.5$ kV/cm for CF_4. The Magboltz simulation (black line) and previous CF_4 data (blue triangle) [32] are also presented.*

that is being adjusted by the signal model $V_{\rm th}(t; v_d, v_{\rm ion}, D_l)$. However, evaluating the likelihood function as a product of likelihoods requires that each $V(t_i)$ are independent of each other. This is obviously not the case as the $V(t)$ signal corresponds to an integration of the induced current and we have checked that the correlation matrix is highly non diagonal. To avoid a diagonalization of the covariance matrix, the time derivate of the signal, $V'(t)$, is used instead. The likelihood function can then be written as the product of the likelihoods associated with each value of $\bar{V}'(t_i)$. It reads as

$$\mathscr{L}(v_d, v_{\rm ion}, D_l, \delta t, A) = \exp\left(-\frac{1}{2}\sum_{i=1}^{N_t}\left[\frac{A \times V'_{\rm th}(t_i - \delta t; v_d, v_{\rm ion}, D_l) - \bar{V}'(t_i)}{\sigma_{\bar{V}'(t_i)}}\right]^2\right) \quad (2)$$

where δt et A are adjusting parameters, to enable a time and amplitude shift between the data and the adjusting model. $\bar{V}'(t)$ is the mean profile value, $\sigma_{\bar{V}'(t)}$ its statistical standard deviation and N_t is the number of time samples. Note that the four nuisance parameters are associated with flat and non informative prior distributions.

4. Experimental results

This method for *in situ* electron drift velocity measurement has been applied to two gas mixtures that might be used for directional Dark Matter detection: pure CF_4 and CF_4 + CHF_3. To examplify, we present on figure 1 (right) the mean profile (black point) and best fit (red curve) for the time derivate of the signal $V'(t)$, for pure CF_4 at 50 mbar and a drift field of $E_d = 137.29$ V/cm. The adjustment is excellent, in particular in the regions of interest for the estimation of the electron drift velocity, *i.e.* rising and falling part of the mean profile $\bar{V}'(t)$. Small differences between the fit and the data in the central region (300 ns $\leq t \leq$ 1200 ns) are due to a lack of accuracy in the estimation of the decreasing part of the preamplifier transfer function.
Figure 2 (left) presents the electron drift velocity measurement in a pure CF_4 gas at 50 mbar, for an amplification field $E_a = 14.5$ kV/cm and a drift field E_d ranging between 50 V/cm and 175 V/cm. The results obtained from a Magboltz simulation [23] are also presented and compared with previous experimental data [32]. Fig. 2 (right) presents the electron drift velocity measurement in a 70% CF_4 + 30% CHF_3 gas mixture at 50 mbar, for an amplification field $E_a = 15.6$ kV/cm. To our knowledge there is no other experimental data with this gas mixture.

As expected, there are discrepancies with the Magboltz simulation as the drift velocity measured accounts for real, but unknown, experimental conditions (impurities, field inhomogeneities, ...) and corresponds to a long drift distance. Note that the electron drift velocity in a 70% CF_4 + 30 % CHF_3 gas mixture is about 5 times lower than in the pure CF_4 case. This is a key point for directional Dark Matter as the track sampling along the drift field will be improved while keeping almost the same Fluorine content of the gas mixture. Fluorine is indeed a golden target [14,15] for spin-dependent Dark Matter search as the spin content is dominated by the unpaired proton [13]. In the case of the MIMAC detector, a fraction of 30% was found to be adequate as it allows to significantly enhance the 3D track reconstruction while conserving sufficiently dense primary electron clouds needed for a high nuclear recoil track detection efficiency.

5. Conclusion

We have presented a new method for *in situ* electron drift velocity measurement, using an alpha source and a profile likelihood analysis based on the modeling of the signal induced on the grid. In particular, we have shown that such analysis allows us to avoid bias due to *e.g.* electron diffusion, ion collection time and electronic readout. Hence, the effective electron drift velocity, *i.e.* in the whole drift space, is measured and it corresponds to the real experimental conditions of the upcoming MIMAC directional detector. We also suggest to add CHF_3 to the standard CF_4 gas used for directional detection as it allows us to lower the electron drift velocity while keeping almost the same Fluorine content of the gas mixture.

References

[1] D. N. Spergel, Phys. Rev. D **37** (1988) 1353
[2] B. Morgan and A. M. Green, Phys. Rev. D **72** (2005) 123501
[3] B. Morgan, A. M. Green and N. J. C. Spooner, Phys. Rev. D **71** (2005) 103507
[4] A. M. Green and B. Morgan, Phys. Rev. D **77** (2008) 027303
[5] A. M. Green and B. Morgan, Astroparticle Physics 27 (2007) 142
[6] A. M. Green and B. Morgan, Phys. Rev. D **81** (2010) 061301
[7] J. Billard, F. Mayet, J. F. Macias-Perez and D. Santos, Phys. Lett. B **691** (2010) 156-162
[8] J. Billard, F. Mayet and D. Santos, Phys. Rev. D **82** (2010) 055011
[9] J. Billard, F. Mayet and D. Santos, Phys. Rev. D **83** (2011) 075002
[10] J. Billard, F. Mayet and D. Santos, Phys. Rev. D **85** (2012) 035006
[11] J. Billard, Q. Riffard, F. Mayet and D. Santos, Phys. Lett. B **718** (2013) 1171
[12] D. S. M. Alves, S. E. Hedri and J. G. Wacker, arXiv:1204.5487 [astro-ph.GA].
[13] D. Albornoz Vásquez, G. Bélanger, J. Billard and F. Mayet, Phys. Rev. D **85** (2012) 055023
[14] S. Ahlen *et al.*, International Journal of Modern Physics A25 (2010) 1-51,
[15] F. Mayet and D. Santos (Eds), *Proceedings of the third workshop on Directional Detection of Dark Matter (CYGNUS 2011) : Aussois, France, June 8-10, 2011*, EAS Publ. Ser. **53** (2012) pp.1.
[16] S. Ahlen *et al.*, Phys. Lett. **B695** (2011) 124-129
[17] E. Daw *et al.*, Astropart.Phys. **35** (2012) 397-401
[18] S. E. Vahsen *et al.*, EAS Publications Series **53** (2012) 43-50
[19] D. Santos *et al.*, EAS Publications Series **53** (2012) 25-31
[20] K. Miuchi *et al.*, Phys. Lett. B **686** (2010) 11
[21] J. Billard, F. Mayet and D. Santos, JCAP **04** (2012) 006
[22] T. Caldwell *et al.*, arXiv:0905.2549
[23] S. F. Biagi, Nucl. Instrum. Meth. **A 421** (1999) 234-240
[24] B. Schmidt, Nucl. Instrum. Meth. **A 252** (1986) 579
[25] B. Schmidt and S. Polenz, Nucl. Instrum. Meth. **A 273** (1988) 488
[26] P. Colas *et al.*, Nucl. Instrum. Meth. **A 478** (2002) 215
[27] F. J. Iguaz *et al.*, JINST **6** (2011) P07002
[28] J. P. Richer *et al.*, JINST **6** (2011) C11016
[29] O. Bourrion *et al.*, EAS Publ. Ser. **53** (2012) 129
[30] J. Billard *et al.*, arXiv:1305.2360
[31] S. Agostinelli *et al.*, Nucl. Instrum. Meth. A **506** (2003) 250.
[32] S. R. Hunter, J. G. Carter and L. G. Christophorou, Phys. Rev. **A38** (1988) 58

Journal of Physics: Conference Series **469** (2013) 012007 doi:10.1088/1742-6596/469/1/012007

Development of a high sensitivity radon detector for purified gases

K. Hosokawa[1], A. Murata[1], Y. Nakano[2], Y. Onishi[1], H. Sekiya[2,3], Y. Takeuchi[1,3] and S. Tasaka[4]

[1] Department of Physics, Kobe University, Kobe, Hyogo 657-8501, Japan
[2] Kamioka Observatory, Institute for Cosmic Ray Research, the University of Tokyo, Higashi-Mozumi, Kamioka, Hida, Gifu, 506-1205, Japan
[3] Kavli Institute for the Physics and Mathematics of the Universe, Todai Institutes for Advanced Study, the University of Tokyo, Kashiwa, Japan 277-8583 (Kavli IPMU, WPI)
[4] Information and multimedia center, Gifu University, Gifu 501-1193, Japan

E-mail: `keishi@stu.kobe-u.ac.jp`

Abstract. For underground particle physics experiments, noble gas radon from U-chain decay series could be a source of serious background events. We have developed a high sensitivity detector for radon using electrostatic collection and PIN photodiode. Recently, we have developed a new type of radon detector. We report basic performance of the new 80L radon detector.

1. Introduction

In underground laboratories around the world, like Kamioka in Japan, LNGS in Italy, SNOLAB in Canada etc, many particle physics experiments are ongoing. Neutrino experiment Super-Kamiokande[1] and dark matter search experiment XMASS[2] are underway at Kamioka. Dark matter search experiment XENON100[3] continues at LNGS. Neutrino experiment SNO+[4] is ongoing at SNOLAB.

All of these experiments need very low background environments, and they are conducted in underground to achieve low background. Especially for underground experiments, noble gas radon from U-chain decay series could be a source of serious background events. Radon gas is generated from decay series of ^{238}U continuously. Because it is contained in the air and has half-life 3.824 days, it exists in large quantities underground. It potentially will dissolve into the xenon and water of XMASS, and in the water of Super-Kamiokande etc. Even if only a little, detector materials contain ^{238}U. The radon concentration has to be reduced in these experiments, and it is necessary accurately know the remaining concentration.

2. High sensitivity radon detector for purified gases

We developed a high sensitivity detector for radon using electrostatic collection and PIN photodiode for underground experiments (70L radon detector)[6]. Radon detectors give the concentration of radon in purified gases or water. In order to detect radon, the radon detector observes alpha ray signals from ^{214}Po decay. The principle of radon detection used in this detector is the electrostatic collection of the daughter nuclei of ^{222}Rn, and the energy measurement of the alpha decay with a PIN photodiode.

2.1. 80L radon radon detector

Figure 1. shows a schematic view of the high sensitivity new 80L radon detector for purified gases. It consists of a stainless steel vessel, a HAMAMATSU-S3204 series 18 mm × 18 mm PIN photodiode, a high voltage divider, amplifier circuit and a feed through. Negative high voltage is supplied to the p-layer of the PIN photodiode, and the vessel is grounded.

Viton O-rings for gaskets and acrylic plates for flanges are used in previous 70L radon detector. They were not airtight. In order to achieve a lower background level, we developed an experimental model of a new detector equipped with airtight parts, copper seal for gaskets and ICF flange. After exchanging the parts, the limit of vacuum level in the detector vessel became 1.0×10^{-5}Pa from 10kPa.

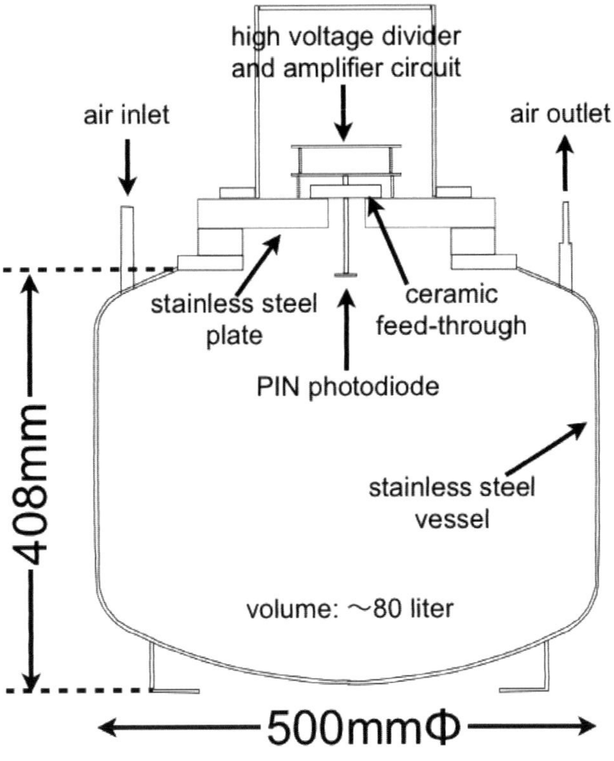

Figure 1. Schematic view of the 80L radon detector

3. Performance evaluation of new 80L radon detector

3.1. Energy spectrum

A typical pulse height spectrum of ^{222}Rn daughter signal is shown in Figure 2.. The ^{214}Po peak signal is used to measure radon concentration, because there are fewer interferering noise events, and no alpha ray peak overlapped its energy region, and it has a higher collection efficiency than ^{218}Po. The signal region of ^{214}Po alpha decay is from 169 to 179 ADC channel for this case. Peak position depends on the supplied voltage value and the settings of the amplifier.

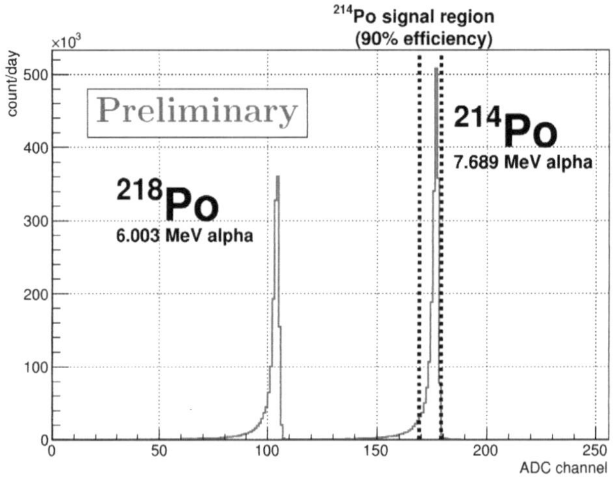

Figure 2. Typical energy spectrum from α decay of ^{222}Rn daughter nucleus

3.2. Calibration factor

To calculate radon concentration in mBq/m^3 from the ^{214}Po count rate in count/day, calibration factor (counts/day)/(mBq/m^3) needs to be obtained. A calibration system was constructed. Using the system, the high voltage dependence and purified gas humidity dependence of the calibration factor were obtained.

3.3. Calibration system

Figure 3. shows a schematic diagram of calibration system for the new 80L radon detector. The detector was connected to a refrigerator to control the dew point temperature, i.e. absolute humidity, a circulation pump, dew point meter, mass flow controller and a ^{222}Rn source. PYLON RNC ^{226}Ra with $0.0783kBq \pm 4\%$ activity was used as ^{222}Rn source. Calibration data was taken with atmospheric pressure purified Ar gas, purified Xe gas or purified air circulation.

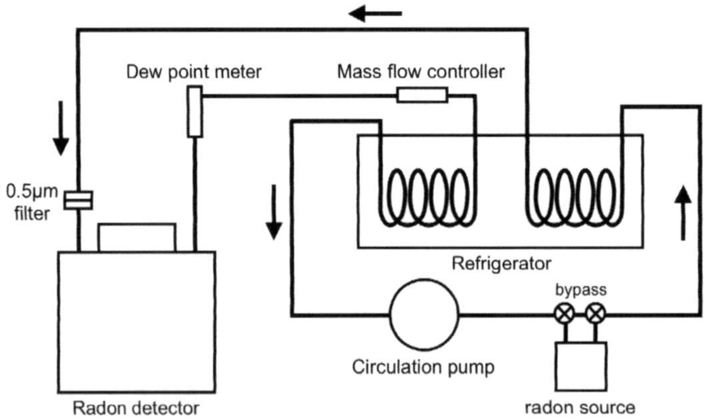

Figure 3. A schematic diagram of calibration system for the new 80L radon detector

3.4. High voltage and humidity dependence of calibration factor

In figure 4., left figure shows calibration factor as a function of high voltage. We can see that the calibration factor of the radon detector rises with higher voltage. The calibration factor became almost settled around -2.0kV.

The right figure shows calibration factor as a function of absolute humidity. Using a refrigerator, we controlled the humidity in order to study the humidity dependence of the calibration factor using the calibration system. Black circle shows old 70L detector's calibration factor with pure air. Colored markers show new 80L detector's calibration factors with pure air, Ar and Xe. New 80L detector with -2.0kV high voltage achieved higher calibration factor than 70L detector. Dotted black line shows 30% detection efficiency in the 80L radon detector. Considering direction of alpha-ray from Po at diode surface, i.e. half of them direct to diode, and signal region covers 90% of ^{214}Po signals, about 70% Po ions are collected by electric field to PIN photo diode surface over dotted line.

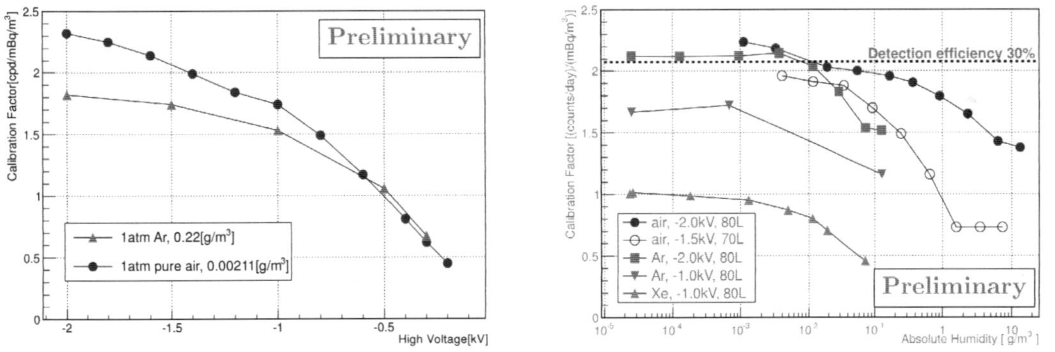

Figure 4. Calibration factor as a function of high voltage(left) and absolute humidity(right)

4. Conclusion

Experimental model of a high sensitivity radon detector was developed and its performance was investigated. Calibration data to calculate its calibration factor were taken. In Kamioka observatory, the radon detectors are used for radon measurement, material screening etc in various underground experiments.

Acknowledgement

This work was supported by JSPS KAKENHI Grant Number 24340050 and ICRR Joint Usage.

References

[1] S. Fukuda et al., Nuclear Instruments and Methods in Physics Research A 501 (2003) 418-462
[2] K. Abe et al., Nucl. Instr. and Meth. in Phys.Res. A661, 50-57 (2012)
[3] E. Aprile et al., Astroparticle Physics 35 (2012) 573-590
[4] C. Kraus, Progress in Particle and Nuclear Physics 57 (2006) 150-152
[5] The Super-Kamiokande Collaboration, Phys.Lett.452(1999)418
[6] Y. Takeuchi et al., Nucl. Instr. and Meth. in Phys. Res. A 421 (1999) 334-341

Energy partition and distribution of excited species in direction-sensitive detectors for WIMP searches

A Hitachi

Kochi Medical School, Nankoku, Japan

E-mail: jm-hitachia@kochi-u.ac.jp

Abstract. The Bragg-like curve for compounds is introduced for directional detection of galactic dark matter. The slow ion collisions are discussed in relation to direct dark matter searches. The Coulomb effect and the threshold effect in stopping power theory are examined. Ionization via molecular orbit (MO) is suggested for an additional contribution to the electronic stopping power at very slow energy.

1. Introduction

Detectors with directional capability, such as gas TPC, can observe daily fluctuations of the WIMP window. The knowledge of the fraction of projectile kinetic energy E into the electronic excitation is a key issue for the direct detection of galactic dark matter. The information on spatial distribution of the electronic energy in the detector medium is essential for the 3D track reconstruction. The Bragg-like curve, $-d\eta/dR_{PRJ}$, where η is the electronic energy given to the target medium and R_{PRJ} is the projection range, has been introduced for practical purpose in gas TPC for directional dark matter searches [1] and may be basically applied also for other directional detection media.

There are criticisms against Lindhard theory [2] that the Coulomb effect and the threshold effect may reduce the electronic stopping power, and consequently the numbers of ionization and excitation produced will be quite small in slow collisions. Also, the recoil nucleus/gamma ratio, RN/γ, ratio may be significantly small at low energy. The stopping mechanism in slow ion collisions will be discussed in relation to those effects. Also the partition of the electronic energy between ionization and excitation in low energy collisions will be considered.

2. Stopping power and LET

Only a part of the energy η goes to the electronic excitation can be used as charge or scintillation signals. The nuclear quenching factor q_{nc} is defined as η/E and was given by Lindhard for recoil ions in single-element media [2] and was later fitted by Lewin and Smith [3]. It is used for quenching calculation for slow recoil ions in condensed phase [4]. For fast particles, contribution to stopping power is exclusively electronic stopping power S_e. Then, the linear energy transfer LET is simply $-dE/dx$. For slow particles the nuclear stopping power S_n becomes the same order of magnitude as S_e. The secondary ions go to collisions again and can give their energy to the electronic excitation. The stopping power, dE/dx, is a quantity belongs to the projectile. The contribution of the secondary electrons and ions are not included. On the other hand, LET is a quantity belongs to the target medium. All the cascade processes has to be counted. The electronic LET ($LET_{el} = -d\eta/dx$), the specific electronic energy deposition along the track of charged particle, can be a good measure for scintillation quenching. For example, Birks low is written as [5],

$$\frac{dL}{dx} = \frac{C_1(-dE/dx)}{1+C_2(-dE/dx)}, \quad (1)$$

for fast particles. However, dE/dx in Eq. (1) should be replaced by dη/dx for slow recoil ions. It is an averaged light yield that is often measured, therefore, an averaged LET_{el}, $< LET_{el} > \approx -\eta/R = -q_{nc}E/R$, where R is the range, may be used for a crude approximation.

^{57}Co 122 keV γ is used as a standard to compare the quenching theory with experimental results. However, the efficiency L_γ for 122 keV γ is necessary to compare measurements with theories in rare gases and inorganic scintillators, since L_γ is less than 1 because of the existence of escaping electrons.

2.1. Nuclear quenching factor for compounds

Detector media for such TPCs are mostly compounds. The charge distribution is given by the distribution of the electronic energy. However, the evaluation of Lindhard factor, the nuclear quenching factor q_{nc}, becomes extremely difficult for the medium contains more than one element. The Lindhard calculation of η mentioned above is valid for $Z_1 = Z_2$, single element media. A simple model, the independent element approach, has been proposed to obtain q_{nc} values for compounds [1]. In binary gases, e.g., taking C ions in C instead of C ions in CF_4, and F ions in F instead of F ions in CF_4. The results calculated for C and F recoils in CF_4 are compared with experimental results in Fig. 1a [6]. The simple model shows a good agreement for F recoils in CF_4. However, the experimental results for C recoils are considerably smaller than the estimated values. One reason may due to the compound ratio [C]:[F] =1:4. C recoils see F atoms 4 out of 5. The Bragg-like curve is given for C and F recoils in CF_4 in Fig. 1b. The axis on the right has been changed using better value of W = 34 eV.

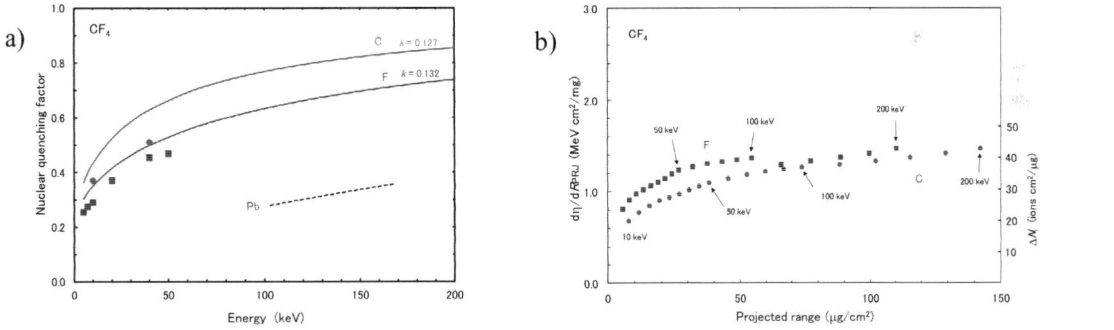

Fig. 1. a) The nuclear quenching factor q_{nc} estimated for C and F recoil ions and recoil Pb ions in α-decay in CF_4, see ref. [1]. The closed circles (C recoil ions) and squares (F recoil ions) are measurements [6]. b) The Bragg-like curve estimated for C and F recoil ions in CF_4. The area below each curve expresses the number of ions produced, N_i. A value of 34.3 eV was taken for the W value for CF_4.

2.2. The energy partition

It has been mentioned that measured q_{nc} values for heavy recoil ions in α-decay in CH_4, C_2H_4 and C_3H_6 seem to be too small [1]. Also, the experimental results for He-He collisions in 5-50 keV was reported about 40-80 % calculated values based on Lindhard model [6]. The result, however, does not necessary mean that calculations by Lindhard model is not adequate. They measured the charge produced in He-C_4H_{10} mixture. The number of ions produced par unit electronic energy deposit for recoil ions can be considerably smaller than those for electrons or α-particles. The average electronic

Journal of Physics: Conference Series **469** (2013) 012008 doi:10.1088/1742-6596/469/1/012008

energy required to produce an electron-ion pair can be different for slow particles in light elements and molecules. The energy balance [7] gives the partition of the electronic energy between the ionization and the excitation,

$$W/I = (\bar{E}_i/I) + (\bar{E}_{ex}/I)(N_{ex}/N_i) + (\bar{\varepsilon}/I), \tag{2}$$

where, I is the ionization potential, N_{ex} and N_i are numbers of excitation and ionization produced, $\bar{E}_{ex}$ and $\bar{E}_i$ are average energy expenditure of excitation and ionization, $\bar{\varepsilon}$ is average energy of sub-excitation electrons. For the fast particles, the optical approximation is good. The resonance states ^1P are excited. The ratio of N_{ex}/N_i can be estimated from the oscillator strength. Eq. (2) gives a constant W/I values for fast particles such as β's and α's within a few % for rare gases. However, ^1S and ^3S states are excited by slow ions and those states are 'true' metastable in He. The energy spent for the metastable states does not give ionization in pure He. When some molecules are introduced in He, the energy transfer from metastable states to the molecules will take place. Those include Penning ionization, however, considerable portion will be wasted for dissociation and non-radiative processes. Triplet states are excited by slow ions also in CH_4, C_2H_4 and C_3H_6 and the energy partition can be different from that by fast ions.

3. Collision theory for slow ions

The Coulomb effect is the deflection and the deceleration of the projectile in the field of the nucleus [8, 9]. The threshold effect states that the energy delivered to the electron must be as large as the ionization energy. We discuss those effects in this section in relation to the heavy ion track, and the molecular orbit (MO) theory.

3.1. Heavy ions

Heavy ions produces co-axial cylindrical track [10]. Direct interaction, the Coulomb interaction, with the primary particle create the core, the inner zone of relatively high-energy density. Bohr's impulse principle considers the incident particle of velocity v interacting only for a duration $2b/v$, where b is the classical impact parameter, defined as the distance of closet approach. The uncertainty between the energy ΔE and the collision time Δt gives:

$$\Delta E \cdot \Delta t \approx \Delta E \cdot \frac{2b}{v} \leq \hbar. \tag{3}$$

If the energy difference of two states E_1-E_0 is less that ΔE, the system cannot be decided between the two states in the collision, then the electronic excitation can take place. Then the radius of the track core b_{max} is given as $b_{max} = \hbar v/2E_1$, where E_1 is the lowest excitation energy. The Coulomb effect prevents the projectile to come close to the target atom in the field of the nucleus when collision velocity v is not fast enough.

The penumbra, the outer zone of low-energy density, is formed by δ-rays produced by close collisions of hard sphere. The maximum energy given to recoil electron is $4(m_e/M)E$, where m_e is the mass of the electron and M is the atom mass, ($m_e \ll M$). The energy of the electron must be higher than E_1 in this simple picture. The experimental evidence shows that the electronic excitation takes place even below the threshold. However, they argue that the threshold effect reduces S_e [8].

Some pointed out [8] that the introduction of the coulomb effect reduces the stopping cross section only by a few % and other stated the reduction is significant [9]. There has been numbers of nuclear stopping power calculation using more refined potentials. They also find that the Lindhard theory gives considerably larger values at low energy for both S_e and S_n. It is the ratio of stopping powers, rather than absolute values, which influences most the nuclear quenching factor $q_{nc} = \eta/E$. The ratio of S_e and S_n by Lindhard theory can be good even if the absolute values are too large. Therefore, the Lindhard theory is still useful at low energy when no other convenient models are available and indeed explains a wide range of experimental results. The Coulomb effect and the threshold effect in the stopping power theory are not important in very low energy since those effects are valid for collisions of low and intermediate, or higher energy. Refinements or corrections of Lindhard theory or Firsov theory are not enough. The real issue in very slow collisions is that separation of the scattering and stopping at very low energy. As Lindhard has pointed out, the nuclear scattering and stopping becomes somewhat uncertain, because the Thomas-Fermi treatment, which most stopping theories for low energy collisions based on, becomes a crude approximation at very low energy [2]. A new approach is necessary as discussed in the following section.

The measurement of stopping power for heavy ions of low velocity is quite difficult. Usually, the total stopping power $S_T = S_e + S_n$ is measured and often Lindhard theory is assumed for either S_n or S_e. Fukuda has measured S_e using a sophisticated apparatus and concluded that Lindhard theory tend to give large S_e values [11]. However, the measurement did not take contributions from $\theta > 0$.

3.2. Molecular orbit

When atomic projectile goes hard (wide deflection angle small impact parameter) collision with atom, the large inelastic energy losses occur at characteristic internuclear distances. Showers of fast-electrons are thrown out [12]. When the collision velocity v_c become much slower than the orbital electron velocity v_e, the electron can follow to the colliding nuclear axis. The transient formation of molecular orbitals (MO's) takes place. The coordinate system for a quasimolecule is composed of an electron and the two nuclei. A significant part of the ion energy can be transferred into the electronic excitation and electrons can be released effectively by very slow ions, such as:

$$\mathrm{Ar}^{+} + \mathrm{Ar} \rightarrow \mathrm{Ar}^{+m} + \mathrm{Ar}^{+n} + (m+n-1)e^{-}. \qquad (4)$$

An electron occupies a MO and become excited during the collision to a higher MO at smaller separation. The system contains vacancies after the collision. Then, energy losses, multiple ionization, ejection of fast electrons and photons follows. Those associated with a large scattering angle θ. The MO theory [13] has been applied mainly to the inner shell excitation and there are no applications for the low energy stopping theory.

References

[1] Hitachi A 2008 *Rad Phys Chem* **77** 1311
[2] Lindhard J *et al* 1963 *Mat Fys Medd Dan Vid Selesk* **33** (10) 1
[3] Lewin J D and Smith P F 1996 *Astropart Phys* **6** 87
[4] Hitachi A 2005 *Astropart Phys* **24** 247
[5] Birks J B 1964 *The Theory and Practice of Scintillation Counting* (Oxford: Pergamnon)
[6] Guillaudin O *et al* 2012 *EAS Pub Ser* **53**, 119
[7] Platzman R L 1961 *Int J Appl Rad Iso* **10** 116
[8] Semrad D 1986 *Phys Rev* A **33** 1646
[9] Tilinin I S 1995 *Phys Rev* A **51** 3058
[10] Mozumder A 1999 *Fundamentals of Radiation Chemistry* (San Diego: Academic Press)
[11] Fukuda A 1981 *J Phys B: At Mol Phys* **14** 4533
[12] Kessel Q C and Everhart E 1966 *Phys Rev* **146** 16
[13] Fano U and Lichten W 1965 *Phys Rev Lett* **14** 627

Journal of Physics: Conference Series **469** (2013) 012009 doi:10.1088/1742-6596/469/1/012009

Measuring Dark Matter Distribution in Directional Direct Detection

Keiko I. Nagao[1]

E-mail: nagao@sci.niihama-nct.ac.jp
Niihama National College of Technology, Department of Engineering Science
7-1 Yagumo Niihama Ehime 792-8580 Japan

Abstract. Direct detection of dark matter with directional sensitivity offers not only measurement of both recoil energy and direction of dark matter, but also a way to understand dark matter distribution in the Galaxy. Maxwell distribution is usually supposed as the distribution near the Earth, however, result of N-body simulations suggest deviation from that due to tidal stream in the Galaxy. We explore the possibility of distinguishing the distribution by direct detection using nuclear emulsions.

1. Introduction

Non-baryonic dark matter is a hot subject for both particle physics and astrophysics. Especially, in particle physics point of view, it can be a valuable clue to the physics beyond the standard model. Many theoretical model which has a candidate for dark matter particle, such as the supersymmetric models and models with extra dimensions, have been proposed.

Direct detection of dark matter imposes severe bounds to such theories, as well as the relic abundance and the indirect detection. It gives lower bound for the interaction cross section between dark matter and nucleon. Event rate R measured in the experiment is converted to the bound for the cross section of dark matter-target atom scattering σ_A by the relation

$$R = N_T n_\chi \int_{E_R} dE_R \int_v d^3v \ f(v) \frac{\sigma_A m_A}{2v\mu_A}, \tag{1}$$

where N_T, n_χ, E_R, v, m_A, and μ_A are the number of target nucleons, the number of dark matter particles, the recoil energy, the velocity of dark matter near the Earth, atomic mass, and reduced mass, respectively. $f(v)$ represents the distribution function of dark matter velocity in the Solar system. Thus modification of the distribution function $f(v)$ can affect the bound for dark matter-nucleon interaction in the theoretical models.

Dark matter distribution in the Galaxy is expected roughly spherical, and we usually adopt Maxwell-Boltzmann distribution

$$f_{\text{Max}}(v) = \frac{1}{(\pi v_0^2)^{3/2}} e^{-(v+v_E)^2/v_0^2}$$

[1] This work is based on a collaboration with Tatsuhiro Naka (Nagoya University) and Mihoko M. Nojiri (KEK and Kavli IPMU).

Journal of Physics: Conference Series **469** (2013) 012009 doi:10.1088/1742-6596/469/1/012009

where v_0 is the velocity of the Solar system, and v_E is the Earth's velocity relative to dark matter. Similar isothermal distributions are also suggested by gravitational potentials of the Galaxy.

However, question whether such distributions is actually realized, has been rose. Halo substructure which can be explained by tidal stream torn off the Sagittarius dwarf galaxy has been reported in the Sloan Digital Sky Survey commissioning data [1]. Cosmological N-body simulation with not only dark matter halo but also baryons, also supports the existence of co-rotating dark halo [2]. Due to the tidal stream, high velocity tail of the dark matter distribution sharply drops compared to Maxwell distribution. Variation from Maxwellian is suggested by debris flow, that is similar to the tidal stream but spatially homogenized, as well.

Direct detection of dark matter with directional sensitivity is next-generation experiment. Measuring both the recoil energy and direction where dark matter particles come from is challenging, and it is more efficient to reject background signals with directional information than ordinary ones. Directional sensitivity will also help to clear the dark matter distribution. In the paper, we study the possibility to measure the local distribution of dark matter focusing on nuclear emulsion detector. Nuclear emulsions have relatively high resolution compared to the gas detectors, therefore suitable to measure the angular distribution of dark matter signals. In Section 2, supposed scattering in our simulation of directional dark matter search with the nuclear emulsions, is introduced. Numerical results are reviewed in Section 3. Conclusions are given in Section 4.

2. Dark matter-nucleus scattering in the nuclear emulsion detector

In the direct search of dark matter, we detect the recoil energy of dark matter-target atom scattering. The nuclear emulsions have several spices for target, mainly, calcium (C), nitrogen (N), oxygen (O), silver (Ag) and bromine (Br). Ag and Br are so heavy that sensitive to relative to heavy dark matter whose mass is $O(100)$ GeV. Light atoms, i.e., C, N and O, are suitable to detect light dark matter, though they are subdominant components accounting for ~30% in weight. In the numerical calculation, we simulate the inelastic scattering of dark matter with mass M_χ GeV to each atom. Mass of the target nucleus is represented as $m_N = 0.932A$ (N =C, N, O, Ag, and Br) and A is mass number.

In the nuclear emulsion detector, the recoil energy is related to the track length left by the dark matter-nucleon scattering. Minimal detectable track length is about 100 nm, which corresponds to ~160 keV (Ag and Br), and ~33 keV (C, N and O) for the recoil energy. We converted the recoil energy to track length in the nuclear emulsions (See [4] for correspondence of the track length and the recoil energy) in the numerical calculation.

In the calculation, z-axis is oriented to the direction of dark matter wind against the Earth, and θ is the scattering angle between z-axis and the track in the nuclear emulsions. Therefore, $\cos\theta = 1$ when the atom is scattered to the counter direction of the Solar system movement in the Galaxy.

3. Numerical Results

Density distributions of the scattering atoms are shown for M_χ =200 GeV in Figure 1, 2 and 3, which correspond dark matter distribution of Maxwellian, with tidal stream, and debris flow, respectively. Since the dark matter velocity is relatively suppressed for Fig. 2 and 3, the track length tends to shorter than that for Maxwell distribution. Similar figures but for M_χ = 800 GeV are shown in Figure 4, 5 and 6. In the calculation, we suppose realistic composition of the nuclear emulsions, in a word, weight ratio of Ag, Br, C, N, and O are 40%, 29%, 12%, 5% and 12%, respectively. Signal tendency for each atom is shown in Appendix, for case of Maxwell distribution.

Journal of Physics: Conference Series **469** (2013) 012009 doi:10.1088/1742-6596/469/1/012009

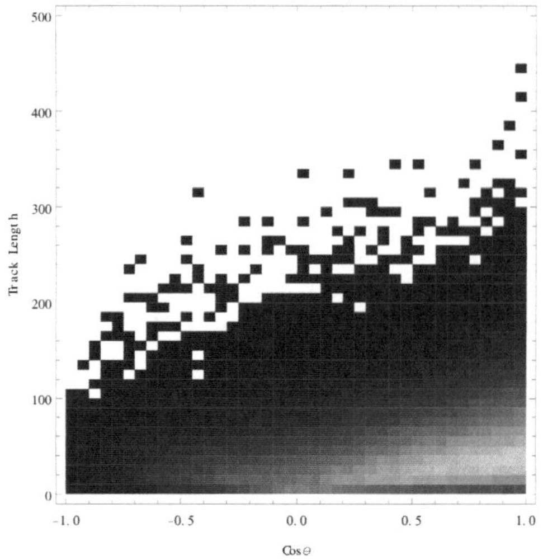

Figure 1. Angular distribution of track length for Maxwell distribution.

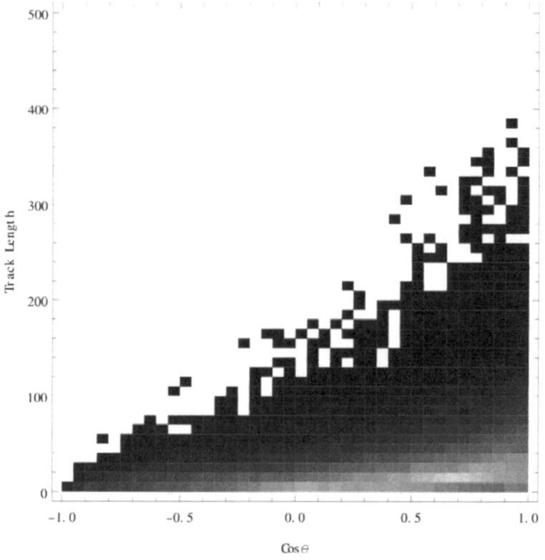

Figure 2. Angular distribution of track length for the dark matter distribution with tidal stream.

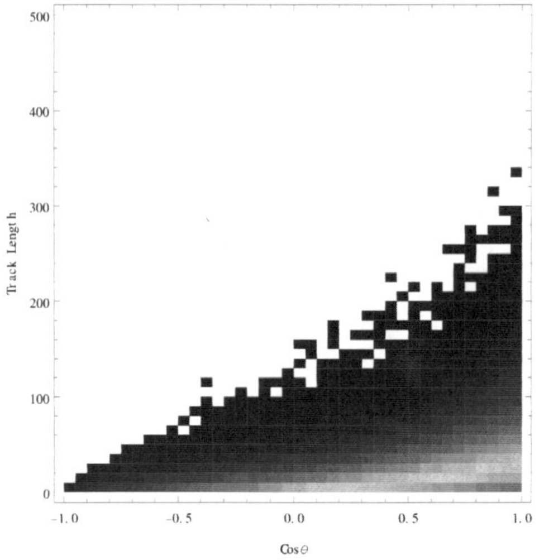

Figure 3. Angular distribution of track length for the dark matter distribution with debris flow.

Density plot of signals of dark matter-nucleon scattering for $M_\chi = 200$ GeV. Horizontal and vertical axes represent the scattering angle $\cos\theta$ and the track length in the nuclear emulsions in nm unit. Yellow (light gray in printed version) region has more signals than brown (dark) one. Figure 1, 2 and 3 correspond to Maxwellian, the dark matter distributions with tidal stream [2] and debris flow [3].

Journal of Physics: Conference Series **469** (2013) 012009 doi:10.1088/1742-6596/469/1/012009

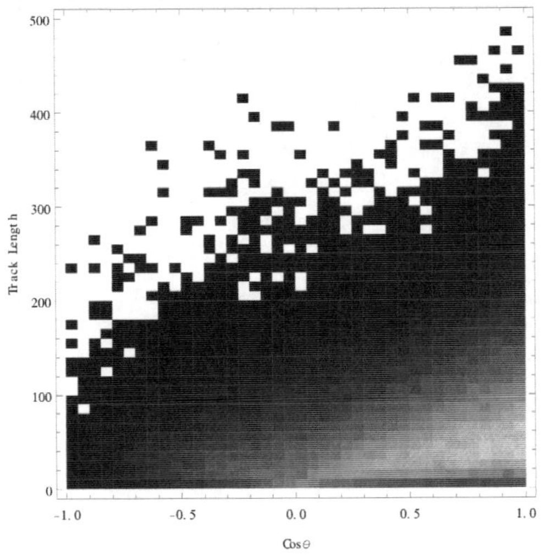

Figure 4. Angular distribution of track length for Maxwell distribution.

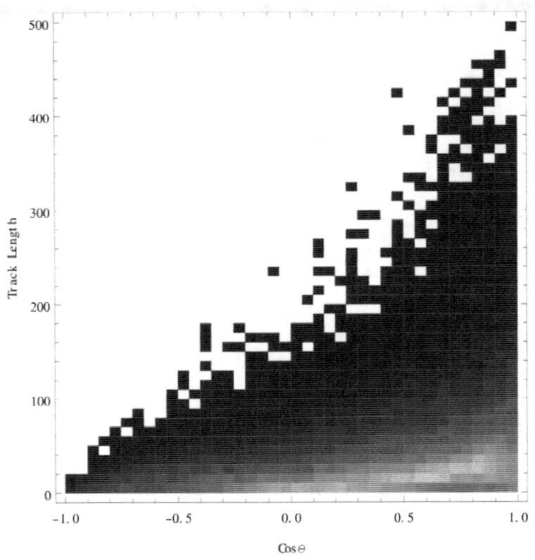

Figure 5. Angular distribution of track length for the dark matter distribution with tidal stream.

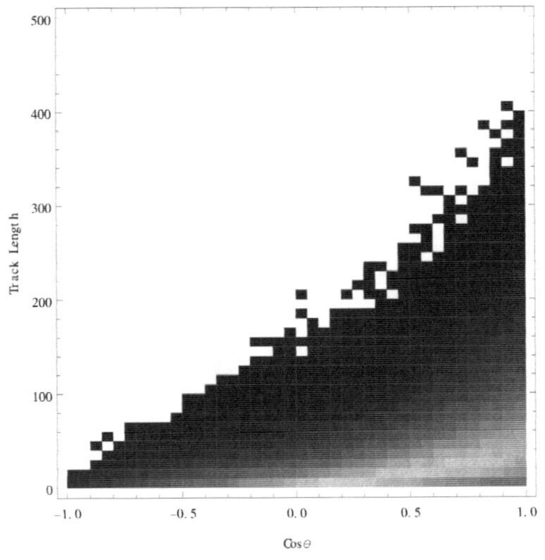

Figure 6. Angular distribution of track length for the dark matter distribution with debris flow.

Legend is same as Figure 1, 2 and 3 except for $M_\chi = 800$ GeV. Heavier dark matter than $M_\chi = 200$ GeV case provides large recoil energy to the scattering atom, and the track length in the nuclear emulsion tends to be long.

Journal of Physics: Conference Series **469** (2013) 012009 doi:10.1088/1742-6596/469/1/012009

4. Conclusions

In this paper, we study the dark matter distribution with direct search of dark matter, focusing on the nuclear emulsion detector. Track and angular distributions are shown for standard Maxwell distribution, and other distribution with tidal stream and debris flow. That of tidal stream has suppressed high velocity tail compared to Maxwell distribution, and also dark matter of the distribution with debris flow has smaller velocity. That makes difference of signal distribution shape even in range where the track is longer than 100 nm, the minimal detectable track length in the nuclear emulsions, if a sufficient number of signals are provided. Dark matter mass is another unknown quantity which can vary the signal distribution, however, it mainly stretches the track length distribution while keeps the signal shape.

References

[1] R. Ibata, M. Irwin, G. F. Lewis and A. Stolte, Astrophys. J. **547**, L133 (2001) [astro-ph/0004255], and references therein.

[2] F. S. Ling, E. Nezri, E. Athanassoula and R. Teyssier, JCAP **1002**, 012 (2010) [arXiv:0909.2028 [astro-ph.GA]].

[3] M. Kuhlen, M. Lisanti and D. N. Spergel, Phys. Rev. D **86**, 063505 (2012) [arXiv:1202.0007 [astro-ph.GA]].

[4] K. I. Nagao and T. Naka, PTEP **2013**, 043B02 (2013) [arXiv:1205.0198 [hep-ph]].

Appendix

Signal distribution for each atom in the nuclear emulsions are shown for Maxwell distribution. We take $M_\chi = 80$ GeV to demonstrate the difference between heavy and light atoms. Ag and Br atoms are much heavier than the other three, it means it is difficult for them to leave long track compared to light atoms.

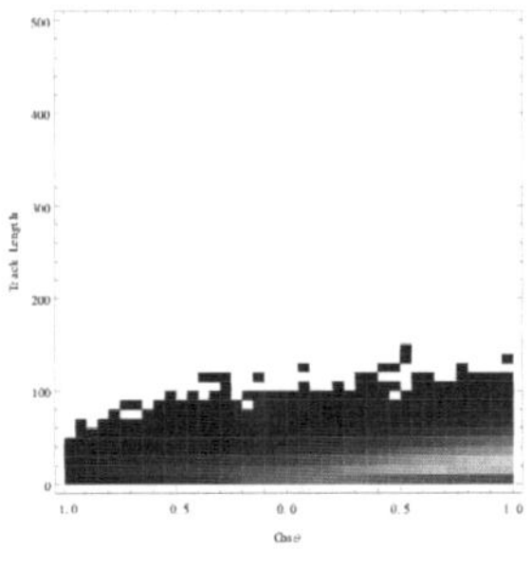

Figure 7. Case for Ag atom

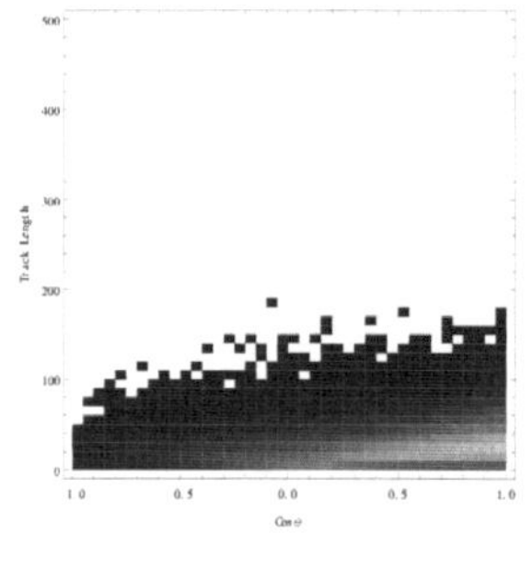

Figure 8. Case for Br atom

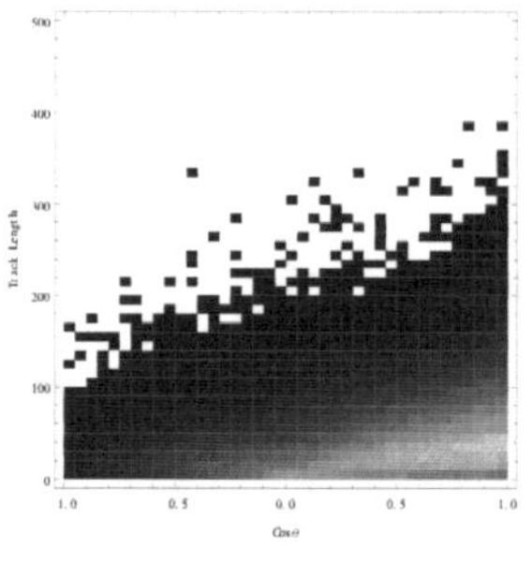

Figure 9. Case for O atom

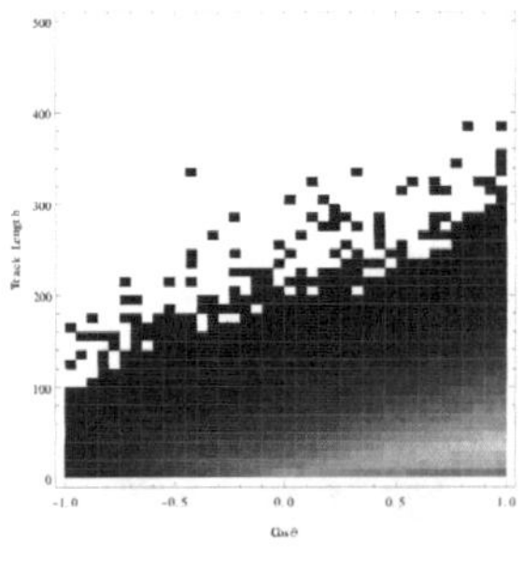

Figure 10. Case for C atom

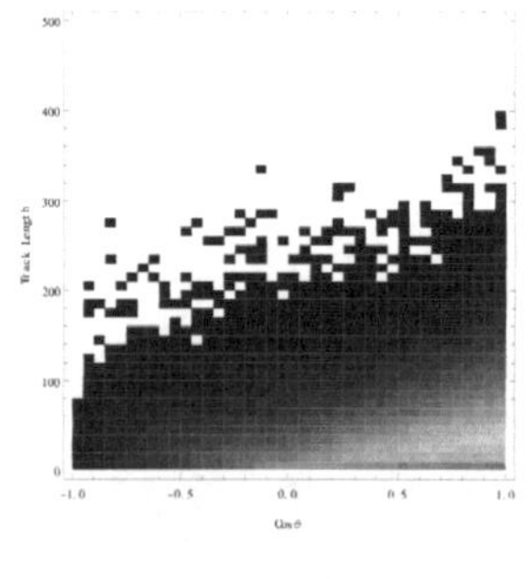

Figure 11. Case for N atom

Legend is same as Figure 1 except for target atoms and $M_\chi = 80$ GeV. Figure 7-11 corresponds to Ag, Br, C, N and O for target atoms.

Journal of Physics: Conference Series **469** (2013) 012010 doi:10.1088/1742-6596/469/1/012010

Detector developing for directional dark matter search with nuclear emulsion

T Asada[1], T Naka[2,3], K. Kuwabara[1], T Katsuragawa[1], M Yoshimoto[1], K Hakamata[1], M Ishikawa[1], M. Nakamura[4], O. Sato[4], and T. Nakano[1]

[1] Graduated School of Science, Nagoya University, Nagoya, Japan
[2] Institute of Advanced Research, Nagoya University, Nagoya, Japan
[3] Kobayashi-Maskawa Institute for The Origin of Particles and The Universe, Nagoya University, Japan
[4] EcoTopia Science Institute, Nagoya University, Nagoya, Japan

E-mail: asada@flab.phys.nagoya-u.ac.jp

Abstract. We are planing the directional dark matter search experiment with nuclear emulsion. Recoiled atoms inside of the emulsion fly several hundred nm, and it is too short to detect with usual emulsion. Fine crystal emulsion was needed to detect such tracks. We developed new method to produce them and succeeded to make crystals small as 20 nm at the minimum size. We also study several methods to improve sensitivity and reduce background noise to survey very interesting cross section region.

1. introduction

Recent results from several experiments indicate detection of WIMP signals. In addition to DAMA/LIBRA [1], Recent result of CoGeNT [3] also suggests annual modulation. Including CRESST excess events [2], their signals were observed at the near mass region from several GeV/c^2 to 100 GeV/c^2 and the near cross sections between 10^{-43} cm^2 and 10^{-39} cm^2. However, some other experiments suggest more lower cross section limit lines. Fore example, that of XENON 100 is two orders lower than the region [4]. In order to conclude this problem, a confirmation from other angle is very important. Directional dark matter search will provide good proof of this problem. We are trying it with nuclear emulsion.

The emulsion is a historical solid detector, and usually used as a film which poured on a plastic support base, as is seen in section on the Figure 1. The emulsion is constructed from two parts, silver halide crystals and dispersion polymer. Each crystal can react to charged particles. When a charged particle passes through the emulsion, it exited electrons. Then exited electrons are captured by a certain spot of the crystal, and small silver grain is generated. A chemical process called developing can supply additional silver to this small grain, and it grow up to visible size. Each grains are lined where the charged particle passed, so finally we can get a track of visible silver grains (Figure 2.).

The flight length of recoiled nucleus from WIMP is estimated to be only several hundred nm. However, it had been proved that fine crystal emulsion can detect such extreme short tracks [5]. Fundamental readout method has also been established with direct expansion technique, and combination of optical microscope and X-ray microscope [6]. In addition to these studies,

Journal of Physics: Conference Series **469** (2013) 012010 doi:10.1088/1742-6596/469/1/012010

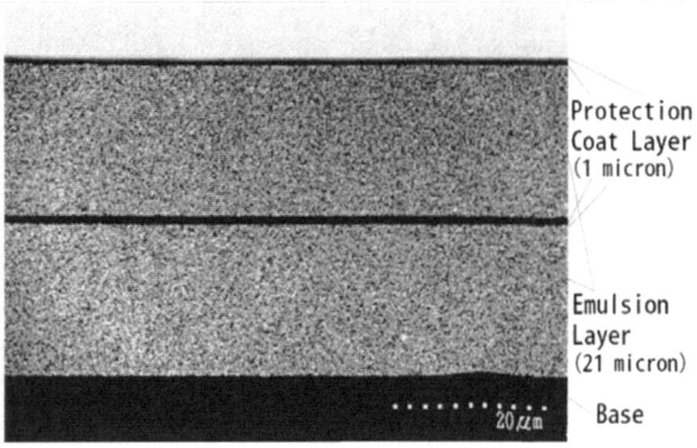

Figure 1. The cross section view of the emulsion film used in OPERA experiment. [7]

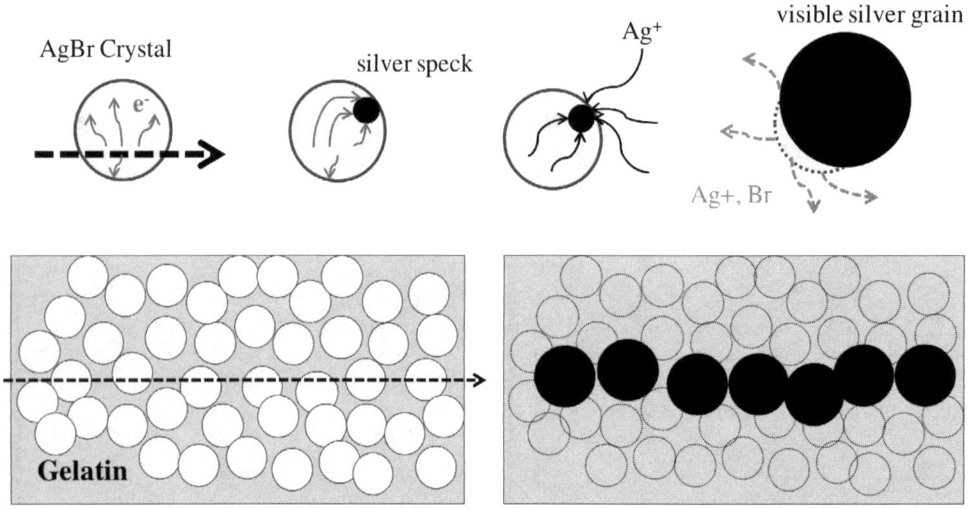

Figure 2. A scheme of reaction for a charged particle passing and developing. Top row describes the process of each crystal. Bottom row is initial and final state of the emulsion.

improvement of resolution, sensitivity and background rejection are necessary to achieve dark matter detection. We have studied how to create our detector itself for these subject.

2. directional dark matter search with nuclear emulsion

The emulsion has several good features for direct dark matter search.

The spatial resolution of the emulsion is quite high. The reason is that dispersion of electrons from ionization is limited inside of each crystals, and initial position can be kept perfectly. Usual emulsion can detect track by micron order resolution, and our special emulsion for dark matter search can catch 100 nm order track.

On the other hand the emulsion does not have time resolution. The direction of WIMP wind is continuously changing by revolution and rotation of the earth, so astronomical information for each event is necessary. Instead of their time record, we can fix the emulsion to equatorial telescope, and follow the direction of WIMP wind.

The emulsion is also suit for the large mass experiment. The emulsion is used as a film, and their readout is same in anywhere. The detector mass affects time for readout linearly, however the readout efficiency is not depend on the mass. Additionally the detector volume will be compressed to relative small size because of solid state.

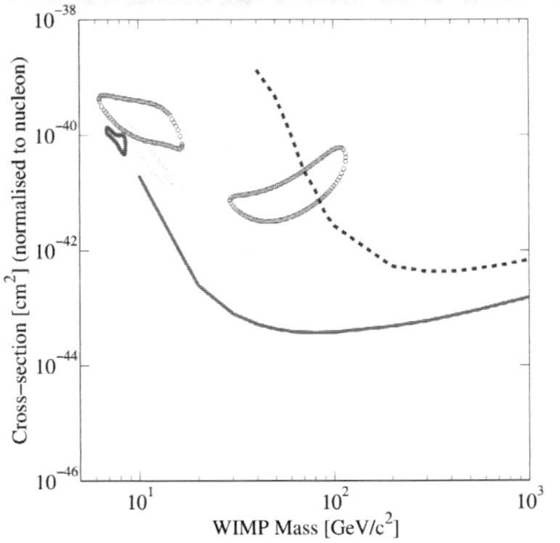

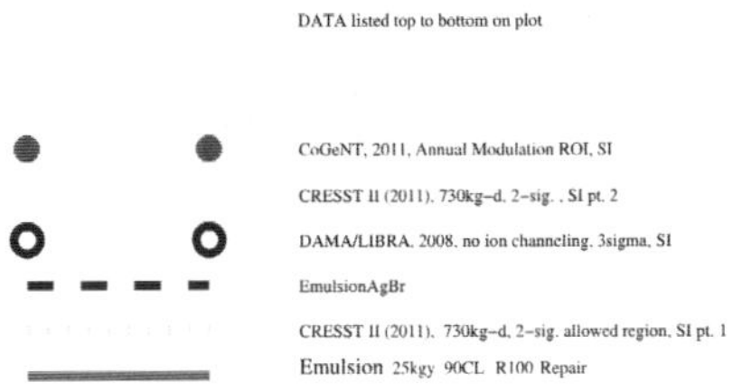

Figure 3. Preliminary simulation of ideal cross section [spin independent, 25 kg·y, 90%C.L.] It are assumed no background and 100% detection efficiency. Blue broken line is only with heavy targets (Ag and Br), and red solid line is including light targets(C, N and O).

3. physics purpose

Our primary purpose is to survey of the interesting region with directional information. Figure 3 shows cross section plot of our experimental sensitivity. It was assumed zero background and 100 % detection efficiency. In principle, our purpose is possible with the resolution which is higher than 100 nm.

Heavy atoms of target may give high crystal sensitivity and effective rejection of backgrounds for their high energy deposit. But most of them cannot fly more than 100 nm and a few events will be detected. To search this region, the detection of light mass atoms like Carbon, Nitrogen, and Oxygen is necessary. Our subject is how to get good sensitivity and background rejection power with these signals while keeping enough resolution. We will reach DAMA region with several ten kg target mass, if we get ideal detector condition.

4. research and development

4.1. production of the emulsion

We started to produce the emulsion by ourselves for rapid research and development. Our emulsion production machine can produce 100 g emulsion per one term in 4 hour. Additionally we are planning new machine which allows three times more rapid production, and we will be able to do at least 100 kg scale of the emulsion experiment.

Important subject for our experiment is how to detect very short tracks like several hundred nm. So first of all, we studied stabilize the size of micronized crystal. Archetype emulsion [5] had 40 nm crystals, and we could also produce it with our new machine. But reproducibility of the size of crystals was not good (Figure4. left).

We tried to make it stable, and find new production method. Polyvinyl alcohol (PVA) can strongly covers surface of crystals and suppress their growing. Usually crystals are generated in gelatin solution, but we added PVA to this solution. This mixing solution makes crystal size very stable even with some additional chemical sensitivity control (Figure 4. right).

4.2. high sensitivity emulsion

We are trying to improve crystal efficiency for carbon track detection. Crystal size is strongly effect to sensitivity, so we have kept the size to relatively large size around 40 nm. When the size is 45 nm, average distance of crystals becomes about 85 nm. Considering our current optical analysis threshold, this size was enough for requirement of test run.

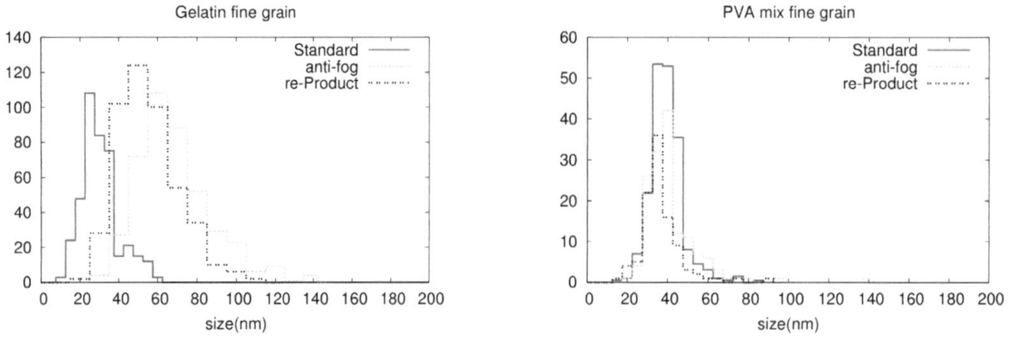

Figure 4. Comparison of crystal size distribution between two emulsion production method. Left is crystal size distribution with only gelatin production method. Right is the one with gelatin and PVA mixed production method.
Red solid line and black dot line are using completely same recipe. Green broken line is added a typical chemical for noise reduction.

Preliminary result have showed that the crystal efficiency for carbon was about 50 % without chemical sensitizing. Further sensitizing study will give very high efficiency for track detection. We started several methods of photographic sensitizing, like reducing recombine of electrons and holes, or disturbing dispersion of silver speck inside of crystal.

4.3. background rejection

Large mass experiments are suffering from many unexpected background signals like β ray from carbon and proton from neutron recoil. We are trying to reject them with crystal sensitivity control. We expect that β rays cannot leave tracks in ideal emulsion. Although protons will be kept sensitivity, we may distinguish it from other target atoms by length and their energy deposit.

The most critical background will be tracks made by chance coincidence. The emulsion generates unexpected silver grains at random with developing (Figure 5). If we define signals as two or more grains within 100 nm sphere, some of them are expected to make fake tracks. Our current emulsion has $0.1 - 1/1000$ um^3 noise grains, so $10^4 - 10^6$ fake tracks will be generated per kg in this case. They will limit experiment of sensitivity.

We are trying to reduce them by two ways. First plan is reduction of unexpected grains. Their amount is proportional to fake tracks by the square. All of Crystal quality, developing, readout efficiency are affect to them. They also affect to sensitivity, so careful study is required. The other way is changing signal definition as more than two grains. Of course signal range threshold will become worse with same crystal quality, however chance coincidence of fake tracks will be greatly reduced. It means the ratio between noise grains and fake tracks are changed from square to cubic. At the case of the definition that three grains are within 200 nm sphere, same noise grains make only $10^{-1} - 10^2$ fake tracks per kg. If $0.1/1000$ um^3 noise grain are assumed, we can run our experiment without this background up to about 10 kg (Figure 6).

4.4. micronized emulsion

If we use 3 grains background rejection, detectable track range becomes longer. Considering energy threshold, we have to reduce distance between crystals. Most effective way is making crystal size smaller. For example, 20 nm crystals have 76 nm of mean distance between 3 crystals, and it is less than the distance between 2 crystals of 45 nm crystals. In fact, we already

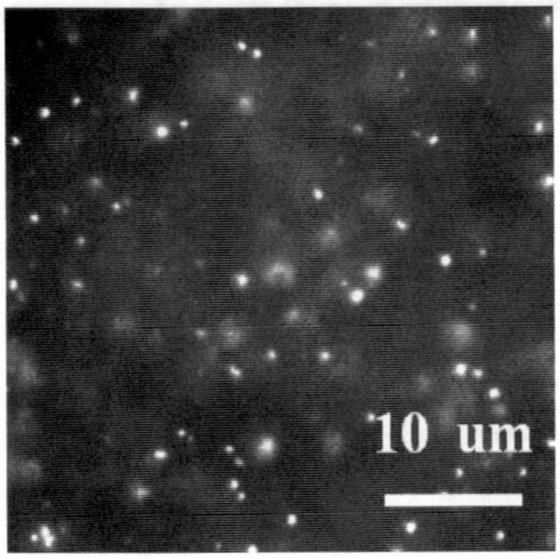

Figure 5. random silver grain in the developed emulsion

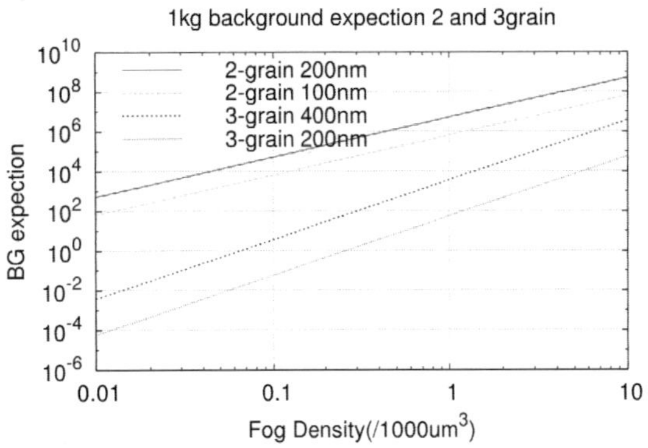

Figure 6. expected number of background tracks by chance coincidence.

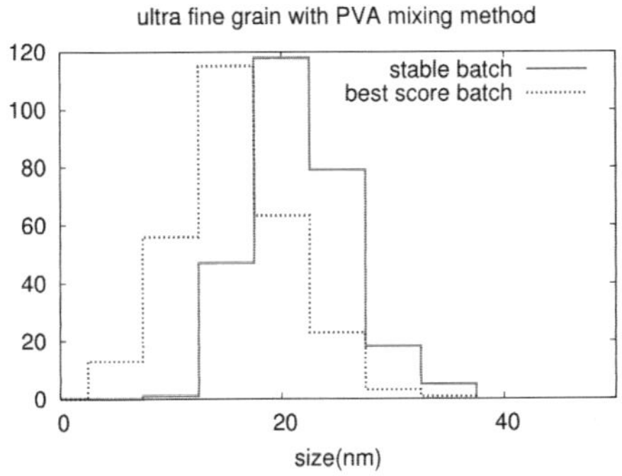

Figure 7. crystal size distribution of quite micronized recipe.
Red solid line is a batch which have good reproducibility, and blue broken line is most smallest batch.

succeeded to make 24 nm stable crystals, and the best one is 18 nm (Figure 7.). These crystal allows further low energy threshold and noise reduction.

Instead of ideal range threshold, smaller crystal has low efficiency. We are planning to make up the loss by strong chemical sensitizing. Because the stability of the crystal size has become quite good, we may be able to sensitize crystals with very strong chemical treatment.

5. conclusion

We developed new fine grained nuclear emulsion to achieve directional dark matter search with large mass experiment. We succeed to produce very stable fine crystals with PVA, and theoretical resolution became enough range. Now we are trying to improve sensitivity and background reduction power. These will lead our experiment to 10 kg target mass order, and very interesting cross section region will be covered.

References

[1] R.Bernabei et al, Eur. Phys. J.,C67, 39 (2010)
[2] G.Angloher et al., Eur. Phys. J C, 72, 1971 (2012)
[3] C.E. Aalseth et al., Phys. Rev. Lett. 107 (2011) 141301
[4] E. Aprile et al, Phys. Rev. Lett. 109, 181301 (2012)
[5] M. Natsume et al., Nucl. Ins. Meth. A, 575 (2007) 439-443
[6] T. Naka et al., Nucl. Ins. Meth. A, 718 (2013) 519-521
[7] Taku Nakamura et al. Nucl. Instr.Meth 556 1(2006) 80

Journal of Physics: Conference Series **469** (2013) 012011 doi:10.1088/1742-6596/469/1/012011

PICO-LON Dark Matter Search

K.Fushimi, S.Nakayama, R.Orito, R.Sugawara and Y.Awatani

Insitiute of Socio- Arts and Sciences, The University of Tokushima, 1-1 Minamijosanjimacho Tokushima city, Tokushima, 770-8502, JAPAN

H.Ejiri and T.Shima

Research Center for Nuclear Physics, Osaka University, 10-1 Mihogaoka Ibaraki city Osaka, 567-0047, JAPAN

R.Hazama

Graduate School and Faculty of Human Environment, Osaka Sangyo University, 3-1-1 Nakagaito, Daito, Osaka, 574-8530, Japan

K,Inoue and H.Ikeda

Research Centre for Neutrino Science, Tohoku University 6-3, Aramaki Aza Aoba,Aobaku Sendai,Miyagi,980-8578,JAPAN

A.Kozlov

Kavli Institute for the Physics and Mathematics of the Universe, 5-1-5 Kashiwanoha, Kashiwa, 277-8583, Japan

E-mail: `kfushimi@tokushima-u.ac.jp`

Abstract. It is an important task to find an annual modulation of WIMPs signal for identifying dark matter signal. To search for annual modulation signal by means of inorganic scintillators, a large mass and a highly radiopure detector is needed. PICO-LON (Planar Inorganic Crystal Observatory for LOw background Neutr(al)inos) is a dark matter search project with multi-array NaI(Tl) scintillators. The test module of PICO-LON whose dimension is 15cm×15cm×0.1cm gives enough performance, energy resolution and low energy threshold, for dark matter search. The highly radiopure NaI(Tl) crystal is indispensable to start a large mass experiment. We have developed highly radiopure NaI(Tl) scintillator. The U and Th chain impurities were precisely measured by meand of pulse shape analysis for NaI(Tl) scintillators. The future plan of KamLAND-PICO dark matter search project is proposed.

1. Introduction

The observation of annual modulation in dark matter signal is the complementary work to directional search for dark matter. The relative velocity of the Earth to the Galactic rest frame is expressed as

$$\vec{v_E} = \vec{v_\odot} + v_{\vec{E}R} \cos\theta, \tag{1}$$

where $\vec{v_E}$ is the Earth's velocity to Galactic rest frame, $\vec{v_\odot}$ and $v_{\vec{E}R}$ are the Sun's velocity to Galactic rest frame and the velocity of the Earth's revolution [1]. The WIMPs (Weakly Interacting Massive Particles), which is the most promising candidate of dark matter [2], has an isotropic directional distribution of velocity to the Galactic rest frame. Consequently, the relative velocity of WIMPS against the Earth modulates with its maximum in June and its minimum in December.

The direction of coming WIMPs velocity cannot be observed by conventional solid state detector, however, the small modulation of energy spectrum is observed by using huge volume detector. DAMA/LIBRA has investigated the annual modulation by using huge highly radiopure NaI(Tl) scintillator, $\simeq$100kg for 7 years and $\simeq$250kg for 6 years. They have reported a significant modulation signal between 2 keV$_{ee}$ and 6 keV$_{ee}$ and no significant modulation above 6keV$_{ee}$, where a subscript *ee* stands for electron equivalent energy[3]. Another experiments with different target reported also the significant annual modulation signals. CDMS and CoGeNT applied Ge detector to search for WIMPs. The cryogenic Ge detector has a great advantage to distinguish a nuclear recoil signal and a β/γ signal. A significant excess in the low energy region which is supposed to WIMPs signal has been reported. However, the regions of WIMPs mass and cross section were completely different.

We propose a dark matter search project using high sensitivity NaI(Tl) scintillator in the northern hemisphere. The huge and highly radiopure NaI(Tl) scintillator will be installed in KamLAND to verify the annual modulation signal.

In the following sections, KamLAND-PICO collaboration will be described. The KamLAND-PICO promotes a high sensitive detection and measurement of WIMPs dark matter. The prototype detector of PICO-LON module and purification of NaI(Tl) crystal will be described.

2. PICO-LON for dark matter detector

PICO-LON stands for Planar Inorganic Crystal Observatory for LOw background Neutralino[6]. Single module of PICO-LON is a thin and wide area NaI(Tl) crystal whose dimension is 0.1 cm in thickness and 10 cm$\sim$15 cm square. The principle of WIMPs selection is shown in Fig.1. A

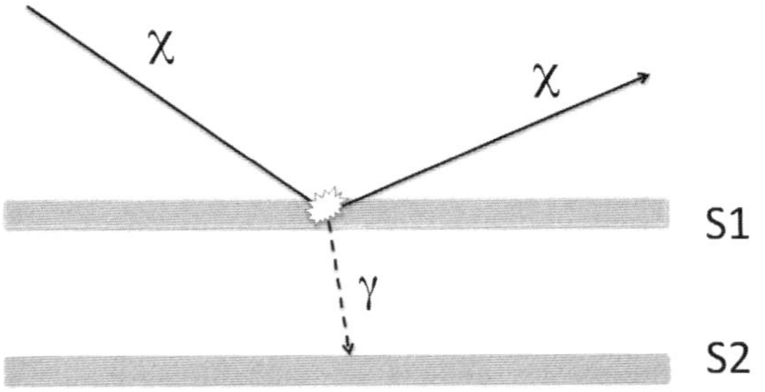

Figure 1. Schematic drawing of the WIMPs detection by inelastic excitation of nucleus.

highly segmented detector system enhances the background reduction power and signal selection power.

The interaction cross section of WIMPs-nucleus scattering is so small that WIMPs can interact with only one module. Background events due to gamma ray and beta ray are rejected by anti-coincidence measurement with neighboring modules.

The another important channel for WIMPs-nucleus interaction is inelastic excitation of nucleus[7]. ^{127}I has a low energy excited state at 57.6 keV which is excited by spin-dependent interaction. The WIMPs with high kinetic energy can easily excite the ^{127}I nuclei and the interaction is followed by immediate radiation of 57.6 keV gamma ray. The experiments using a large volume detectors [8, 9] observe the total energy deposition which is the sum of recoil energy and gamma ray energy. The inelastic signal consequently shift to higher energy region. The large amount of background suffers the sensitivity to inelastic excitation for normal large volume detectors.

PICO-LON detector system enables to make a coincidence measurement of inelastic excitation signal. The thickness of the scintillator is as thin as 0.1 cm. Thus the gamma ray escapes from the interaction module (S1) and is detected by the adjacent module (S2). Then the nuclear recoil signal from S1 is measured in coincidence with the gamma ray signal from S2.

The performance of a prototype module of PICO-LON was tested. The energy resolution and the energy threshold were measured by irradiating low energy gamma rays. The scintillation photons were collected from four thinner edges by photomultiplier tubes (PMT) through a lightguide. The lightguide which was made of acrylic plastic acts as airtight container as well. Three modules of PMTs whose diameter is 2 inches were glued on each thinner edge of light guide. The PMT outputs from four edges were individually input into analog-to-digital converter.

The energy spectrum taken by PICO-LON single module which was irradiated by a ^{241}Am gamma ray source is shown in Fig.2 The good energy resolution was obtained by using

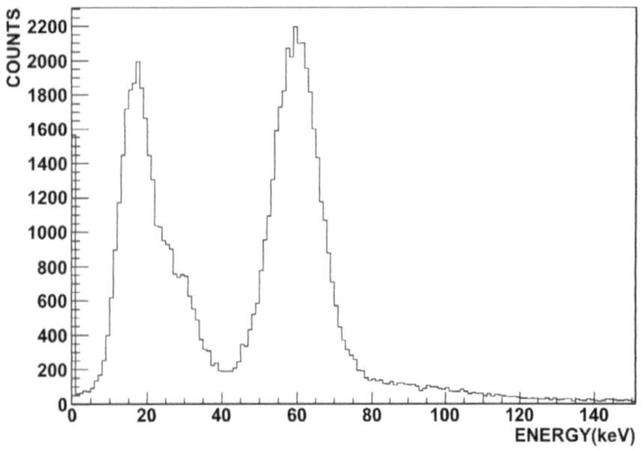

Figure 2. The energy spectrum obtained by PICO-LON module by irradiationg ^{241}Am source.

uncollimated irradiation. The energy resolution at 59.5 keV γ ray was 14 keV in full-width-half-maximum. The energy threshold of the detector was limited by electrical noise which was mainly from PMT dark current and base line fluctuation. The obatined energy threshold was as low as 2 keV electron equivalent energy. Both the energy resolution and the energy threshold were good enough to search for WIMPs[10].

Purification of NaI(Tl) crystal is the most important task for dark matter search. DAMA group developed extremely radiopure NaI(Tl) crystals [11]; the concentration of U-chain and Th-chain impurities are less than 10 ppt. Japanese companies are also developing high purity NaI(Tl) crystals. Recently, we restarted to develop highly radiopure NaI(Tl) crystal. Present status of impurities in NaI(Tl) scintillator reported by DAMA and the present work are shown in table 1. One of the most serious impurities is ^{210}Pb which makes various types of background against dark matter search. In the present work, a sufficiently small concentration of ^{210}Pb was obtained by the Japanese maker.

3. KamLAND-PICO for dark matter search

Kamioka Liquid Scintillator Anti-Neutrino Detector (KamLAND) is the huge neutrino detector which is located in Kamioka underground laboratory in Japan [12]. The center of the detector is 1 kton of ultrapure liquid scintillator (LS) contained in 13-m-diameter spherical balloon made of 135 μm-thick transparent nylon/EVOH (ethylene vinyl alcohol copolymer) composite film.

Table 1. Impurities of NaI(Tl) reported by DAMA and our present crystal. The goal of PICO-LON project is also shown.

Impurity	DAMA	Present Crystal	Goal of PICO-LON
^{40}K	< 20 ppb	Not yet	< 20 ppb
Th chain	0.5~0.7 ppt	< 1 ppt	< 1 ppt
U chain	0.7~10 ppt	8 ppt	< 1 ppt
^{210}Pb	5~30 μBq/kg	50 μBq/kg	< 100 μBq/kg

The radioactive impurities in LS are sufficiently small for dark matter search. The concentrations of U-chain and Th-chain was $(3.5 \pm 0.5) \times 10^{-18}$ g/g and $(5.2 \pm 0.8) \times 10^{-17}$ g/g, respectively. The impurity of ^{40}K is less than 2.7×10^{-16} g/g, which is sufficiently low for dark matter search.

KamLAND-PICO is the project which searches for WIMPs dark matter by installing PICO-LON detector into KamLAND. The final phase of KamLAND-PICO consists of 1 ton of multilayer NaI(Tl) scintillator and the huge LS tank. First phase of KamLAND-PICO detector consists a simple bulk NaI(Tl) scintillator whose total mass is 1 ton. The first phase of KamLAND-PICO aims to find WIMPs signal using ultra pure NaI(Tl) crystal. The final phase of KamLAND-PICO aims to determine the property of WIMPs interaction.

4. Conclusion

PICO-LON has shown enough performance of low energy threshold (~2 keV) and good energy resolution (23% FWHM at 60 keV). The low background which is the most important property has been established by selecting surrounding materials. The purity of NaI(Tl) ingot is as low as 8 ppt for U-chain and less than 1 ppt for Th-chain contaminants. Further improvement of impurity and stability test is in progress.

The sensitivity to WIMPs in the first phase of KamLAND-PICO is enhanced by a large volume detector and high selectivity using an efficient active shield. The expected sensitivity for spin-independent WIMPs reaches to the order of 10^{-9} pb for 40~100 GeV/c^2.

References

[1] K.Freese, J.Freeman and A.Gould, Phys. Rev. **D37** (1988) 3388.
[2] G.Jungman, M.Kamionkowski and K.Griest, Phys. Rep. **267** (1996) 195.
[3] R.Bernabei et al., arXiv:1301.6243.
[4] R.Agnese et al., arXiv:1304.4279.
[5] C.E.Aalseth et al., Phys. Rev. Lett.**106** (2011) 131301.
[6] K.Fushimi et al.,J. Phys. Soc. Jpn. **74** (2005) 3117.
[7] J.Ellis, R.A.Flores and J.D.Lewin, Phys. Lett. **B212** (1988) 375.
[8] H.Ejiri, H.Ohsumi and K.Fushimi, Phys. Lett. **B317** (1993) 14.
[9] R.Bernabei et al., New Jour. Phys. **2** (2000) 15.1.
[10] K.Harada et al., KEK Proceedings of the 26th Workshop on Radiation Detectors and Their Uses, (2012) 122.
[11] R.Bernabei et al., Nucl. Instru. & Meth. in Phys. Res. **A592** (2008) 297.
[12] K.Eguchi et al., Phys. Rev. Lett. **90** (2003) 021802.

Journal of Physics: Conference Series **469** (2013) 012012 doi:10.1088/1742-6596/469/1/012012

Status of R&D on double phase Argon detector: the ANKOK project

Masashi Tanaka
Waseda University, Japan
E-mail: masashi.tanaka@aoni.waseda.jp

Abstract. Liquid argon is known to be an excellent target material for WIMP dark matter direct search experiment. Since PY2012, we are working on a new dark matter search project (ANKOK) at Waseda university using the double phase argon detector technique. As a first stage of the project, we are currently concentrating our effort on improvement of the scintillation light detection efficiency so that the experiment has better sensitivity for low mass down to 10 GeV/c^2 WIMP. In this report, we describe R&D status of the ANKOK experiment.

1. Introduction

Liquid argon is known to be an excellent target material for WIMP (weakly interacting massive particle) dark matter direct search experiment. Use of its ionization and scintillation signals, and scintillation pulse shape provides strong discrimination ($>10^8$) between the electron and nuclear recoil events. Since liquid argon is inexpensive, it has a large advantage for the future large (ton scale) experiments. Relatively small atomic mass (A=40) gives higher nuclear recoil energy for WIMP-Ar nuclear scattering. Thus argon has potential advantage for low mass WIMP ($\sim$10 GeV/c^2) search. On the other hand, the 128 nm VUV scintillation light of argon is relatively hard to detect with nominal photo sensors, and use of wavelength shifter reduces the light detection efficiency, $\sim$5 pes/keVee for argon [1, 2] compared to $\sim$15 pes/keVee xenon [3].

Since PY2012, we are working on a new dark matter search project (ANKOK) at Waseda university using double phase argon detector technique. We are currently concentrating our effort on further improvement of the scintillation light detection efficiency so that the experiment has better sensitivity for low mass ($\sim$10 GeV/c^2) WIMP. In the next few years, we are targeting to construct a detector with fiducial mass of several tens of kg, and to collect the underground physics data.

In January 2013, we have built a double phase argon detector with fiducial mass of 10 kg. Target of this study is to establish basic techniques, such as safety handling of the cryogenic liquid, purification of liquid argon, and stable operation of the double phase detector. Selected results on this study are described in section 3.

In August 2013, we have built a simple single phase liquid argon detector with fiducial mass of 0.2 kg. Target of this study is to maximize the light collection efficiency of the argon scintillation light, and to quantitatively understand the background discrimination ability of pulse shape discrimination, which are described in section 4.

2. Waseda Liquid Argon Test Stand

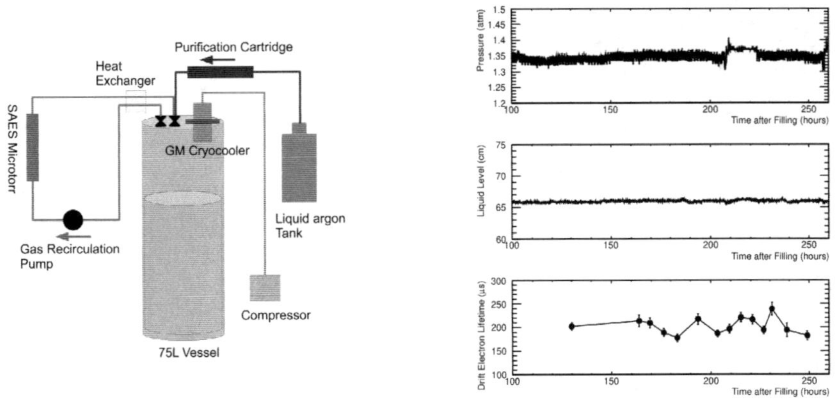

Figure 1. Schematic view of Waseda liquid argon test setup(left), and monitoring plots for 10 days of detector operation, inner vessel pressure (right-top), liquid surface level (right-middle), liquid argon purity (right-bottom).

Waseda liquid argon test stand is located at Nishi-Waseda campus of Waseda university. Figure 1 shows the schematic view of the cryogenic and argon purification system. The cryostat has its size of about 30 cm diameter and 100 cm length (75L vessel). A Gifford-MacMahon(GM) cryocooler (Sumitomo CH-110) with available cooling power of 200W (@90K) is mounted on top of the vessel. A molecular turbo pump (Pfeiffer Hipace-300) is also mounted on top of the vessel and the vessel is evacuated to $\sim 10^{-4}$ Pa for 2 weeks to remove impurities inside the vessel. Purification of liquid argon is firstly performed during filling the vessel. Commercial liquid argon which typically contains impurity about 1 ppm is passed through a hand-made purification cartridge [4]. Additionally a gas recirculation and purification system (SAES Microtorr) has been implemented to maintain and improve the liquid argon purity. We note that both the purification cartridges have ability to remove electronegativities (oxygen, water, etc) but not for nitrogen impurities which can be affected to scintillation light yield.

Figure 1 shows monitoring summary plots for $\sim$10 days of data taking run. During this period, we have successfully operated the detector: Inner vessel pressure has kept 1.35 ± 0.1 atm; liquid surface level has kept within 1mm; attenuation time of drift electron has been stable at $\sim$200 μs, which corresponds to $\sim$1.5 ppb O_2 equivalent impurity.

3. Results from 10 kg Double Phase Detector

The 10 kg double phase detector has radius of 25 cm and drift length of 20 cm. Maximum -20 kV (1 kV/cm) of high voltage is generated by using 20-stage Cockcroft-Walton (CW) generator located inside the liquid argon, and each stage of CW generator is connected to field shaping ring made by 1/8" stainless-steel tube. Figure 2 shows picture of the 10 kg double phase detector from the backview of CW location. Argon scintillation light is detected by using 5 PMTs located bottom of the detector, 4 of 2" PMT (HAMAMATSU R6041-506) and 1 of 3" PMT (HAMAMATSU R11065). These PMTs do not have quantum efficiency for 128 nm argon scintillation light, so surface of PMT window is coated by TPB wavelength shifter (128 nm to 430 nm). Quantum efficiencies for 2" PMT and 3" PMT are $\sim$ 25% and $\sim$30%, respectively at 430 nm. For simplicity, in this detector we do not locate reflectors at the side of the detector, thus the geometric acceptance for scintillation light collection is small (<3%). Top of the detector is 1 mm-thick copper anode plate, and a stainless-steel grid plate with 100 um thickness and 4 mm

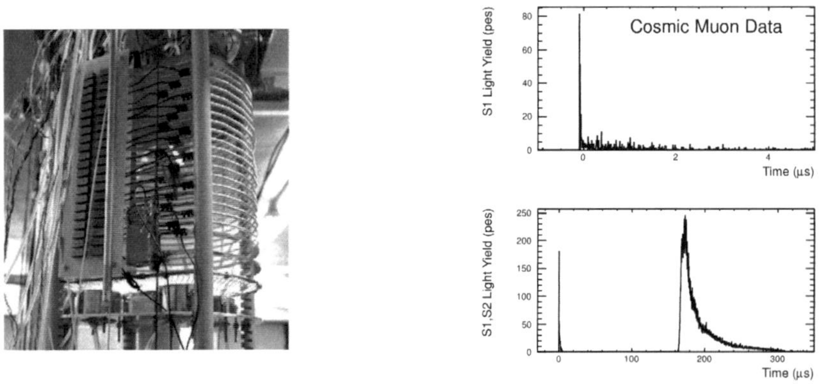

Figure 2. Picture of the 10 kg double phase detector (left), and typical electron recoil event (right-top) found in cosmic triggered data (right-bottom).

pitch is inserted 1 cm bellow the anode plate. Liquid surface is kept in the middle of the anode plate and the grid plate, and $\sim$ 3 kV/cm of electric field is supplied for drift electron extraction. The PMT signal is read out through 100MS/s 14bit FADC (CAEN V1724). We have collected electron recoil event using cosmic muon triggered by external hodoscope. Figure 2 shows typical electron recoil event. This event has significant slow component with time constant of $\sim$1.5 μs (top plot) in S1 signal, and large S2 peak (bottom plot) which is due to proportional light for extracted electrons from liquid to gas.

4. Results from Single Phase Detector

Since target of the 0.2 kg single phase detector is to optimize the light collection efficiency, we have build a simple and small detector. It is consists of 2 of 2" PMT located face to face with 6 cm distance, and TPB coated reflector (3M ESR) surrounding these PMTs (Fig. 3). We have collected electron recoil events using γ-rays from several radioactive sources, ^{57}Co, ^{60}Co, ^{22}Na, ^{137}Cs, and ^{252}Cf. Also collected nuclear recoil events using neutrons from ^{252}Cf source.

Figure 3 shows scintillation light signal waveform for nuclear recoil (top) and electron recoil (bottom) events found in ^{252}Cf data sample. While nuclear recoil event has small fraction of slow component, electron recoil event has large slow component as expected.

4.1. Light Yield

We have determined light yield of the 0.2 kg detector in the unit of the energy deposition of γ rays (pes/keVee). Figure 4 shows observed light yield distribution for ^{137}Cs γ ray (=662 keV). Points and histogram in the figure correspond to the data and simulation, respectively. The nominal light yield is determined by scaling the horizontal axis of the simulated distribution so that the simulation and data distribution are the best χ^2 match. ^{137}Cs γ ray sample indicates 1.90$\pm$0.12 pes at the best χ^2 value. The same procedure is performed for other γ sources and the result summarized is in Fig. 4. All data points are consistent with $\sim$1.8 pes/keVee within a standard deviation.

Following studies for increasing the light yield are ongoing. (1) 2" PMT (HAMAMATSU R60401-506) we have used for the 0.2 kg detector has relatively small quantum efficiency (25% at 430 nm) while we have PMT (HAMAMATSU R11065-20) with higher quantum efficiency (38% at 430 nm) in hand. (2) Further optimization of the TPB coated reflector (3) Improvement in liquid argon purity especially removing the nitrogen impurities. (4) Use of infrared argon

Journal of Physics: Conference Series **469** (2013) 012012 doi:10.1088/1742-6596/469/1/012012

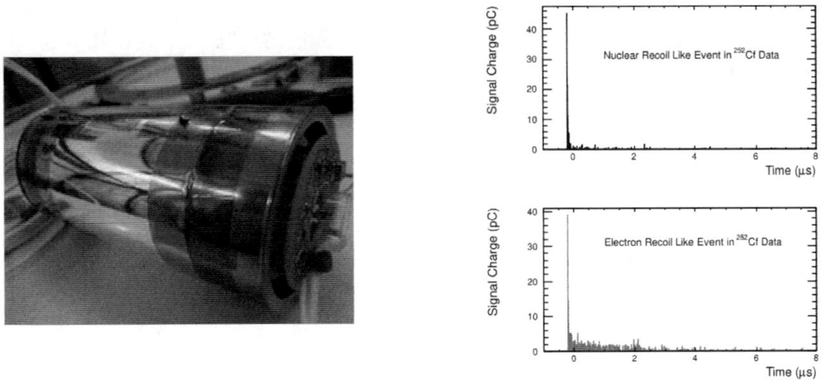

Figure 3. Picture of the 0.2 kg single phase detector (left), and typical nuclear recoil (right-top) and electron (right-bottom) recoil events in ^{252}Cf data sample.

scintillation light [5].

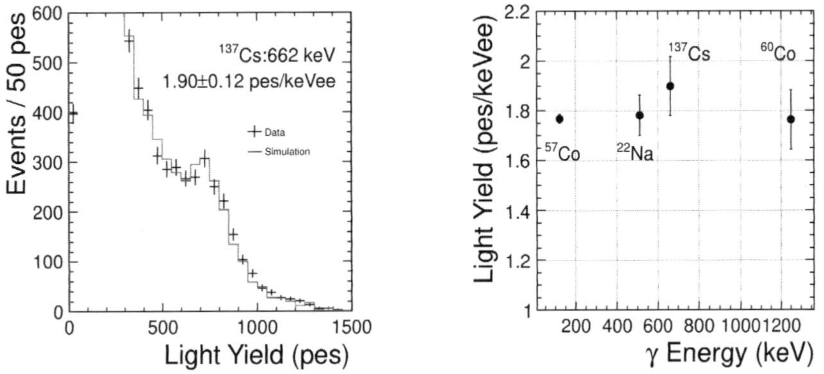

Figure 4. Observed number of photoelectron ^{137}Cs data sample in the 0.2 kg detector (left), and summary of light yield measured by using several γ sources (right).

4.2. Background Reduction

To estimate background reduction ability of the 0.2 kg detector, we have define a simple parameter, Slow/Total, which is fraction of the light yield observed as a slow component to total light yield. Total and Slow light yields are defined as collected light yields in the time range of [-50 ns, 8 μs] and [100 ns, 8 μs], respectively while t=0 corresponds to the trigger timing. Figure 5 shows the Slow/Total distribution for ^{252}Cf data sample. One can find clear two peaks which can be well distinguished, the larger peak around Slow/Total = 0.7 is the electron recoil events and the other peak around Slow/Total = 0.2 corresponds to nuclear recoil events. Note that the distribution is made after selecting the event with Npe > 100 pes, the separation power between two peaks is rather vague for the events with lower light yield.

The shape of the Slow/Total distribution can be predicted by simple binomial distribution $P[X = k] = {}_nC_k p^k (1-p)^{n-k}$ while n is the number of observed photoelectron, and p is the average number of Slow/Total. Figure 5 shows the Slow/Total distribution in ^{22}Na data (points) and binomial model (histogram) for the case on n is around 500 pes. Data and binomial model are in good agreement, but the data shows small excess in both the lower and higher tail. Studies to understand features of such tail events are in progress.

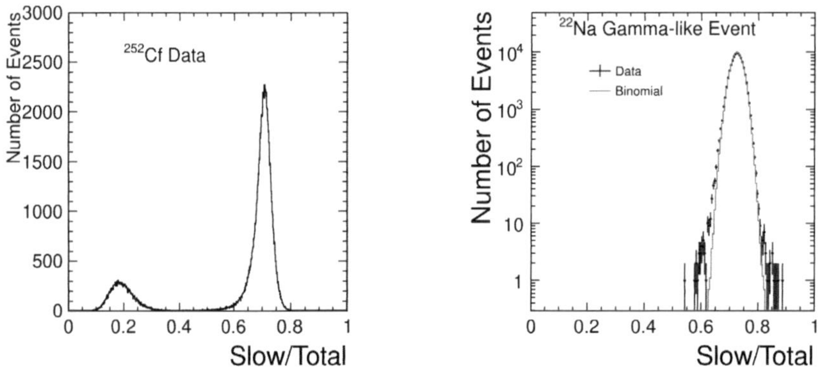

Figure 5. Slow/Total distribution for ^{252}Cf data sample (left). Comparison of Slow/Total distribution in data (right-points) and binomial expectation (right-histogram).

5. Summary

In PY2012, we have started a new dark matter search project (ANKOK) at Waseda university using double phase argon detector technique. We have established basic techniques for the double phase argon detector operation by using the 10 kg double phase detector. To maximize the argon scintillation light collection efficiency, we have build the 0.2 kg single phase detector, and so far, we have achieved ~1.8 pes/keVee of light yield. We still have several handles to improve the light yield and such studies are ongoing. Electron and nuclear recoil type events are collected using several radioactive sources, and we have studied background reduction ability by defining simple parameter, Slow/Total. We found the Slow/Total distribution was well modeled by simple binomial model. More detailed and quantitative studies to understand the background reduction ability are ongoing.

Acknowledgments

This experiment is a part of the outcome of research performed under a Waseda University Research Institute for Science and Engineering (Project number 13C09). This work was supported by JSPS Grant-in-Aid for Challenging Exploratory Research Grant Number 25610060.

References

[1] M. G. Boulay, B. Cai, M. Chen, V. V. Golovko, P. Harvey, R. Mathew, J. J. Lidgard and A. B. McDonald *et al.*, arXiv:0904.2930 [astro-ph.IM].
[2] D. Akimov *et al.* [DarkSide Collaboration], arXiv:1204.6218 [astro-ph.IM].
[3] K. Abe, K. Hieda, K. Hiraide, S. Hirano, Y. Kishimoto, K. Kobayashi, S. Moriyama and K. Nakagawa *et al.*, Nucl. Instrum. Meth. A **716**, 78 (2013) [arXiv:1301.2815 [physics.ins-det]].
[4] A. Curioni, B. T. Fleming, W. Jaskierny, C. Kendziora, J. Krider, S. Pordes, M. P. Soderberg and J. Spitz *et al.*, Nucl. Instrum. Meth. A **605**, 306 (2009) [arXiv:0903.2066 [physics.ins-det]].
[5] A. Bondar, A. Buzulutskov, A. Dolgov, A. Grebenuk, E. Shemyakina and A. Sokolov, JINST **7**, P06015 (2012) [arXiv:1204.0180 [physics.ins-det]].

Journal of Physics: Conference Series **469** (2013) 012013 doi:10.1088/1742-6596/469/1/012013

A review on the discovery reach of Dark Matter directional detection

F. Mayet[1] **and J. Billard**[2,3]

[1] Laboratoire de Physique Subatomique et de Cosmologie, Université Joseph Fourier Grenoble 1, CNRS/IN2P3, Institut Polytechnique de Grenoble, 53, rue des Martyrs, Grenoble, France
[2] Department of Physics, Massachusetts Institute of Technology, Cambridge, MA 02139, USA
[3] MIT Kavli Institute for Astrophysics and Space Research, Massachusetts Institute of Technology, Cambridge, MA 02139, USA

E-mail: mayet@lpsc.in2p3.fr

Abstract. Directional detection of galactic Dark Matter offers a unique opportunity to identify Weakly Interacting Massive Particle (WIMP) events as such. Depending on the unknown WIMP-nucleon cross section, directional detection may be used to : exclude Dark Matter, discover galactic Dark Matter with a high significance or constrain WIMP and halo properties. We review the discovery reach of Dark Matter directional detection.

Several projects of directional detectors [1–9] are being developed for Dark Matter search. There is in particular a worldwide effort toward the development of a large TPC devoted to this goal [1–7]. Since the pionner paper of D. N. Spergel [10], the contribution of directional detection to the field of Dark Matter has been adressed through a wealth of studies [11–34]. Depending on the unknown WIMP-nucleon cross section, directional detection may be used to : exclude Dark Matter [13, 17], discover galactic Dark Matter with a high significance [12, 15, 25] or constrain WIMP and halo properties [14, 26, 27].

1. Directional framework

Dark Matter directional detection aims at measuring both the energy (E_r) and the 3D track (Ω_r) of a recoiling nucleus following a WIMP (Weakly Interacting Massive Particle) scattering. The double-differential spectrum is given by

$$\frac{d^2R}{dE_r d\Omega_r} = \frac{\rho_0}{4\pi m_\chi m_r^2}\left[\sigma_0^{SI} F_{SI}^2(E_r) + \sigma_0^{SD} F_{SD}^2(E_r)\right]\hat{f}(v_{min}, \hat{r}) \tag{1}$$

where m_χ is the WIMP mass, ρ_0 the local WIMP density, m_r the reduced WIMP-nucleus mass, σ_0^{SI} (resp. σ_0^{SD}) the spin independent (resp. dependent) WIMP-nucleus cross section at zero momentum transfer, F_{SI} (resp. F_{SD}) the spin independent (resp. dependent) form factor, v_{min} the WIMP minimal velocity to produce a recoil and $\hat{f}$ the three-dimensional Radon transform of the WIMP velocity distribution $f(\vec{v})$, given by [34]

$$\hat{f}(v_{min}, \hat{q}) = \int \delta(v_{min} - \vec{v}.\hat{q}) f(\vec{v})\, d^3v \tag{2}$$

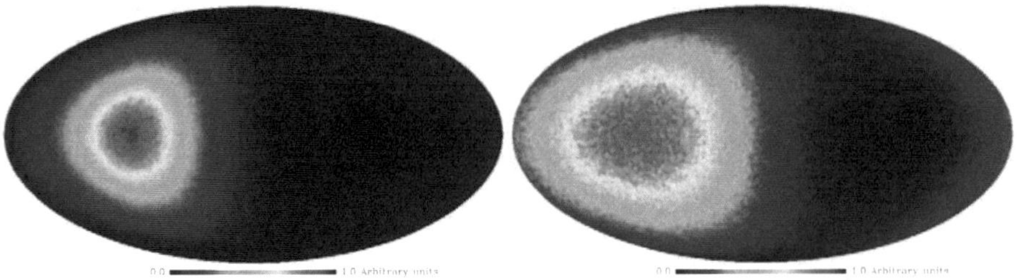

Figure 1. (Left) : WIMP flux for an isothermal spherical halo. (Right) WIMP-induced recoil distribution. Recoils maps are produced for a ^{19}F target, a 100 GeV.c^{-2} WIMP and considering recoil energies in the range 5 keV $\leq E_R \leq$ 50 keV. Figures extracted from [12].

The expression of the WIMP velocity distribution $f(\vec{v})$ depends on the Milky Way halo model. For an isotropic isothermal sphere, it reads

$$f(\vec{v}) = \frac{1}{(2\pi\sigma_v^2)^{3/2}} \exp\left(-\frac{(\vec{v}+\vec{v}_\odot)^2}{2\sigma_v^2}\right) \tag{3}$$

where $\vec{v}_\odot$ is the Sun velocity vector and σ_v is the WIMP velocity dispersion.
It follows that the WIMP event distribution is expected to present an excess in the direction of motion of the Solar system ($-\vec{v}_\odot$), which happens to be roughly in the direction of the constellation Cygnus ($\ell_\odot = 90°, b_\odot = 0°$ in galactic coordinates). As shown in [12], the WIMP-induced recoil distribution presents a dipole-feature (fig. 1) while the background distribution [35] is expected to be isotropic in the galactic rest frame. In fact, several directional features provide a clear and unambiguous difference between the WIMP signal and the background one, *e.g.* dipole [12], ring-like[1] [28], aberration[2] [29] and daily modulation of the WIMP direction.
The event spectrum (1) depends on the particle model (m_χ, σ_0^{SI} and σ_0^{SD}) and on the Dark Matter halo model ($\rho_0, f(\vec{v})$). For direction-insensitive Dark matter search (dR/dE_r), this high number of free parameters may induce a bias due to wrong halo model assumption when constraining the WIMP properties (mass and cross section), see *e.g.* [36]. Thanks to the measurement of the double-differential spectrum (1), directional detection may either account for astrophysical uncertainties [15,16] or even contrain astrophysical parameters [14,26,27].
Low pressure TPC cannot be arbitrarily large and are hence exposure-limited. This is the reason why most directional detectors are using a target made of a light nucleus with non-vanishing spin, that makes them sensitive to the spin dependent interaction (σ_0^{SD}) for which current limits are weaker. As shown in [12], all target nuclei present[3] an equivalent directional signal, when adjusting the energy range. For low pressure TPC, ^{19}F is usually considered as the golden target for SD directional detection.

2. Dark Matter exclusion

Two dedicated exclusion methods have been developed for directional detection [13, 17]. S. Henderson *et al.* have proposed [17] a 2D generalization of the maximum gap [37], first proposed for direction-insensitive Dark Matter detection in order to deal with an unknown background contamination. It is based on the double-differential spectrum (1) and allows to account for all information given by a directional detector. However, the energy spectrum of the

[1] maximum of the recoil rate in a ring around the mean recoil direction.
[2] annual variation of the mean recoil direction.
[3] at sufficiently low recoil energy when the form factor can be approximated to unity

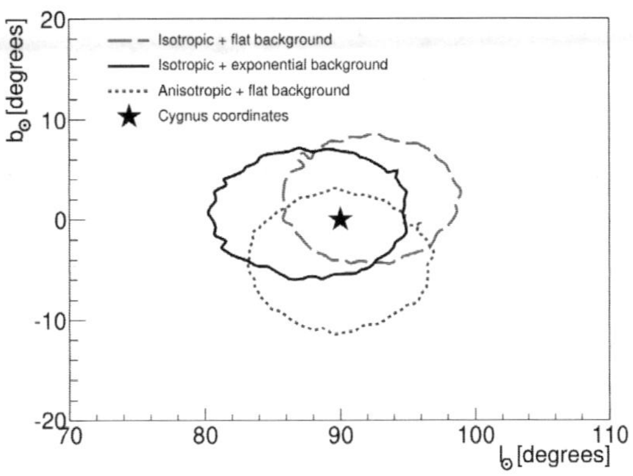

Figure 2. 95% contour level in the $(\ell_{\odot}, b_{\odot})$ plan for three input models: Isotropic halo model + exponential background (solid line), Isotropic halo model + flat background (long dashed line) and anisotropic halo model + flat background (dotted line). Figure is extracted from [14].

background is unknown, while its angular spectrum is expected to be isotropic in the galactic rest frame. J. Billard *et al.* have taken advantage on this point [13] by proposing a likelihood method that deals only with the angular part of the spectrum in a given recoil energy range. This way, no assumption on the background energy dependence is needed and conservative exclusion limits may be provided. As shown in [13] a 30 kg.year CF_4 directional detector would be able to exclude spin dependent Dark Matter down to $\sim 10^{-6}$ pb in the background free case and $\sim 10^{-5}$ pb with a background rate of 10 events/kg/year (with sense recognition).

3. Dark Matter discovery

Beyond the exclusion strategy, directional detection may be used to discover Dark Matter [12, 15, 25]. First, one may prove that the directional data are not compatible with the background, by rejecting the isotropy hypothesis. With the help of unbinned likelihood methods [22] or non-parametric statistical tests on unbinned data [24], it has been shown that a few number of events $\mathcal{O}(10)$ is required to reject the isotropy hypothesis. These methods [18] are based on a generic test of isotropy following the mean recoil deviation $<\cos\theta>$. For instance, the significance of an observed anisotropy can be evaluated by computing the distributions of $<\cos\theta>$ for both H_0 corresponding to the isotropic background hypothesis and H_1 the background plus signal hypothesis. The use of the variable $<\cos\theta>$ is particularly interesting in the case of directional detection of Dark Matter as the expected signal exhibits a dipole feature [12] hence maximizing the deviation between H_0 and H_1.

One may also show that the data favor the background plus signal hypothesis (H_1). The method proposed in [12] is a blind likelihood analysis, that allows to identify a genuine WIMP signal as such. The proof of discovery is the fact that the recovered main recoil direction is pointing towards Cygnus, within a few degrees at 95% CL (see fig. 2), even with a sizeable background contamination and in the case of non standard Dark Matter halo models. This outlines the robustness of this parameter as an observable to prove a positive detection of Dark Matter with a directional detector. Even at low exposure, a high significance discovery is achievable for various detector configurations [15]. Moreover, it is possible to go beyond the standard Dark Matter halo paradigm [15] by accounting for most astrophysical uncertainties [38]. This is a key advantage for directional detection with respect to direction-insensitive strategy. Indeed, as the velocity dispersions are set as free parameters, induced bias due to wrong model assumption

Journal of Physics: Conference Series **469** (2013) 012013 doi:10.1088/1742-6596/469/1/012013

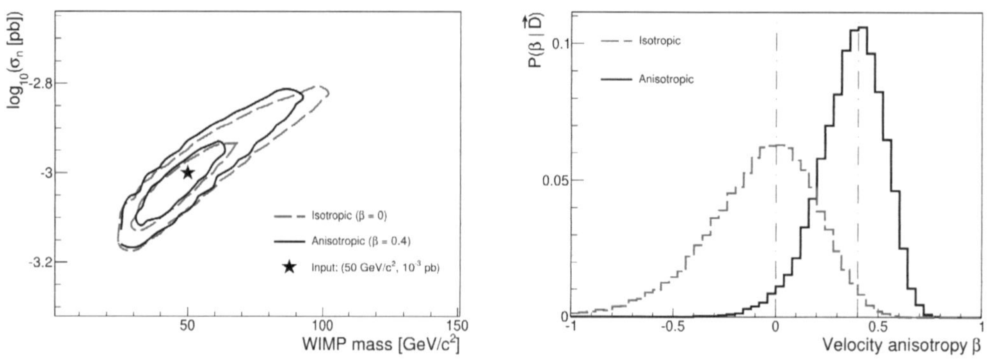

Figure 3. Left panel : 68% and 95% contour level in the (m_χ, σ_n) plane, for a 50 GeV/c^2 WIMP and for two input models : isotropic ($\beta = 0$) and triaxial ($\beta = 0.4$). Right panel : posterior PDF distribution of the β parameter for the same models. Figures are extracted from [14].

should be avoided. This is for instance the effect observed in [36], with a systematic downward shift of the estimated cross section, when assuming a standard isotropic velocity distribution fitting model whereas the input model is a triaxial one [39].

4. Dark Matter identification

For high WIMP-nucleon cross section, it is also possible to go further by constraining the WIMP and halo properties [14] thanks to a high dimensional multivariate analysis. Indeed, a 30 kg.year CF_4 directional detector would allow us to constrain the WIMP properties, both from particle physics (mass and cross section) and galactic halo (velocity dispersions). Figure 3 presents the constrains on m_χ, σ_n and β, the velocity anisotropy parameter, that may be obtained with a single 30 kg.year directional detector. Hence, directional detection may allow to constrain models beyond the standard model of particle physics as well as to discriminate between various halo models.

In a so-called *post-discovery era*, meaning the WIMP mass is supposed to be known to sufficient precision, it has been shown that directional detection may be used to infer Dark Matter phase space distribution in the solar neighborhood [26]. In particular, a parametrization of the functional form of the Dark Matter distribution is proposed, avoiding to rely on ansatzes. In this case, the coefficients of its moment decomposition on a model independent basis are the measurable quantities in a directional experiment. The conclusion of [26] is that about 1000 events are required for a good measurement of the underlying Dark Matter distribution.

5. Constraining Dark Matter and supersymmetry models

As shown in [11], directional detection provides a powerful tool to explore neutralino Dark Matter models as most MSSM configurations, and to a lesser extent for NMSSM ones, with a neutralino lighter than 200 GeV/c^2 would lead to a significance greater than 3σ (90% CL) in a 30 kg.year CF_4 directional detector. No signal with such an exposure would lead to an exclusion of neutralinos up to 600 GeV/c^2.

The use of directional detection to constrain the astrophysical properties of Dark Matter has received much interest in the past years. Beyond the constraint on the Dark Matter halo properties [14] (*e.g.* velocity dispersions of the local WIMP velocity distribution), directional detection may be sensitive to the presence of substructures in the Milky Way halo, such as Dark Matter tidal streams (spatially localized), debris flow (spatially homogenized but with velocity substructures) and a co-rotating dark disk. Such components of the local Dark Matter distribution may lead to distinctive features in the expected directional signal [16, 27, 31, 32, 40], although the conclusion depends strongly of their unknown properties. As a matter of fact, constraining their properties remains however a challenging task requiring a very low threshold and/or a large exposure, depending on the type of substructure.

6. Conclusion

Dark Matter directional detectors with large exposure ($\sim$ 30 kg.years) offer a unique opportunity as they may lead, depending on the value of the unknown WIMP-nucleon cross section, either to a conclusive exclusion, a high significance discovery of galactic Dark Matter or even to an estimation of the WIMP properties. For larger exposures, directional detection may be a way to break the neutrino floor that stands as the ultimate limit for direct Dark Matter detection [41].

References

[1] S. Ahlen *et al.*, Int. J. Mod. Phys. A **25** (2010) 1
[2] F. Mayet and D. Santos (Eds), *Proceedings of the third workshop on Directional Detection of Dark Matter (CYGNUS 2011) : Aussois, France, June 8-10, 2011*, EAS Publ. Ser. **53** (2012) pp.1.
[3] S. Ahlen *et al.*, Phys. Lett. **B695** (2011) 124-129
[4] E. Daw *et al.*, Astropart.Phys. **35** (2012) 397-401
[5] S. E. Vahsen *et al.*, EAS Publications Series **53** (2012) 43-50
[6] D. Santos *et al.*, EAS Publications Series **53** (2012) 25-31
[7] K. Miuchi *et al.*, Phys. Lett. B **686** (2010) 11
[8] A. Drukier, K. Freese, D. Spergel, C. Cantor, G. Church and T. Sano, arXiv:1206.6809
[9] T. Naka *et al.*, EAS Publ. Ser. **53** (2012) 51
[10] D. N. Spergel, Phys. Rev. D **37** (1988) 1353
[11] D. Albornoz Vásquez, G. Bélanger, J. Billard and F. Mayet, Phys. Rev. D **85** (2012) 055023
[12] J. Billard, F. Mayet, J. F. Macias-Perez and D. Santos, Phys. Lett. B **691** (2010) 156-162
[13] J. Billard, F. Mayet and D. Santos, Phys. Rev. D **82** (2010) 055011
[14] J. Billard, F. Mayet and D. Santos, Phys. Rev. D **83** (2011) 075002
[15] J. Billard, F. Mayet and D. Santos, Phys. Rev. D **85** (2012) 035006
[16] J. Billard, Q. Riffard, F. Mayet and D. Santos, Phys. Lett. B **718** (2013) 1171
[17] S. Henderson, J. Monroe and P. Fisher, Phys. Rev. D **78** (2008) 015020
[18] B. Morgan and A. M. Green, Phys. Rev. D **72** (2005) 123501
[19] B. Morgan, A. M. Green and N. J. C. Spooner, Phys. Rev. D **71** (2005) 103507
[20] C. J. Copi and L. M. Krauss,, Phys. Rev. D **63** (2001) 043507
[21] C. J. Copi *et al.*, Phys. Rev. D **75** (2007) 023514,
[22] C. J. Copi and L. M. Krauss, Phys. Lett. B **461** (1999) 43
[23] A. M. Green and B. Morgan, Phys. Rev. D **77** (2008) 027303
[24] A. M. Green and B. Morgan, Astroparticle Physics 27 (2007) 142
[25] A. M. Green and B. Morgan, Phys. Rev. D **81** (2010) 061301
[26] D. S. M. Alves, S. E. Hedri and J. G. Wacker, arXiv:1204.5487 [astro-ph.GA].
[27] S. K. Lee and A. H. G. Peter, JCAP **1204** (2012) 029
[28] N. Bozorgnia, G. B. Gelmini and P. Gondolo, JCAP **1206** (2012) 037
[29] N. Bozorgnia, G. B. Gelmini and P. Gondolo, JCAP **1208** (2012) 011
[30] R. J. Creswick, S. Nussinov and F. T. Avignone, Astropart. Phys. **35** (2011) 62
[31] M. Kuhlen, M. Lisanti and D. N. Spergel, arXiv:1202.0007 [astro-ph.GA].
[32] M. Lisanti and J. G. Wacker, Phys. Rev. D **81** (2010) 096005
[33] M. S. Alenazi and P. Gondolo, Phys. Rev. D **77** (2008) 043532
[34] P. Gondolo, Phys. Rev. D **66** (2002) 103513
[35] D. Mei and A. Hime, Phys. Rev. D **73** (2006) 053004
[36] A. M. Green, JCAP **0708** (2007) 022
[37] S. Yellin, Phys. Rev. D **66** (2002) 032005, and arXiv:0709.2701
[38] A. M. Green, Mod. Phys. Lett. A **27** (2012) 1230004
[39] N. W. Evans, C. M. Carollo and P. T. de Zeeuw, Mon. Not. Roy. Astron. Soc. **318** (2000) 1131
[40] A. M. Green, JCAP **1010** (2010) 034
[41] J. Billard, L. Strigari and E. Figueroa-Feliciano, arXiv:1307.5458

AUTHOR INDEX

Institute of Physics
Dirac House, Temple Back
Bristol BS1 6BE UK

CURRAN ASSOCIATES INC
proceedings.com